Frederick James Britten

On the Springing and Adjusting of Watches

Frederick James Britten

On the Springing and Adjusting of Watches

ISBN/EAN: 9783337365530

Printed in Europe, USA, Canada, Australia, Japan

Cover: Foto ©berggeist007 / pixelio.de

More available books at **www.hansebooks.com**

ON THE

Springing

AND

Adjusting of Watches

Being a description of the Balance Spring and the Compensation Balance with Directions for applying the Spring and Adjusting for Isochronism and Temperature.

BY

F. J. BRITTEN,

Author of the Watch & Clock Makers' Handbook, Former Clock and Watch Makers, &c.

London:

E. & F. N. SPON, 125, STRAND, W.C.

New York:

SPON & CHAMBERLAIN, 12, Cortlandt Street.

PREFATORY NOTE.

To meet many enquiries for a book of moderate price explanatory of the springing and adjusting of watches, this little work is submitted.

It is intended for those tolerably conversant with watch-making generally, yet who desire guidance in this particular branch, rather than for beginners, and therefore a knowledge of many elementary facts is assumed.

The examination of modern watch escapements is dealt with because it belongs particularly to the art of the adjuster. For drawings of the escapements, particulars of their action, and other details, the student is referred to the "Watch and Clock Makers' Handbook," to which this volume may be regarded as supplementary.

A brief historical notice of the balance spring and compensation balance is included here; a comprehensive account of the early craftsmen may be found in "Former Clock and Watch Makers and their Work."

If springs and balances of nickel-steel alloy answer all expectations it may be that, in the future, adjustment for varying temperatures will be unnecessary. But in such an event so many springs and balances of other material will remain as to justify the inclusion of those pages devoted to compensation, even apart from their historical interest.

F. J. B.

35, Northampton Square,
      London, E.C.

CONTENTS.

CHAPTER I.
Introduction and Effect of Various Springs.

CHAPTER II.
Theoretically Correct Terminal Curves.—Effect of Disturbing Influences.

CHAPTER III.
Compensation for Varying Temperature.

CHAPTER IV.
Method of Procedure in Springing and Adjusting.

CHAPTER V.
The Manufacture of Balance Springs.

CHAPTER VI.
Non-magnetic Material and Material insensible to Changes of Temperature.

CHAPTER VII.
Gauges.

CHAPTER VIII.
Observatory Tests, and Note on Timing Repeating Carriage Clocks.

CHAPTER IX.
Examination of Escapements.—Revolving Escapements.

ON THE
SPRINGING AND ADJUSTING OF WATCHES.

CHAPTER I.

The vibrating wheel of a watch or chronometer which, in conjunction with the balance spring, regulates the progress of the hands is called the balance. The time in which a balance will vibrate cannot be predicated from its dimensions alone. A pendulum of a given length always vibrates in the same time as long as it is kept at the same distance from the centre of the earth, because gravity, the force that impels it, is always the same; but the want of constancy in the force of the balance spring, that in watches and chronometers takes the place of gravity, and governs the vibrations of the balance is one of the chief difficulties of the timer. There is another point of difference between the pendulum and the balance. The time of vibration of the former is unaffected by its mass, because every increment of mass carries with it a proportional addition to the influence of gravity: but by adding to the mass of a balance, the strength of the balance spring is not increased at all, and therefore the vibrations of the balance become slower.

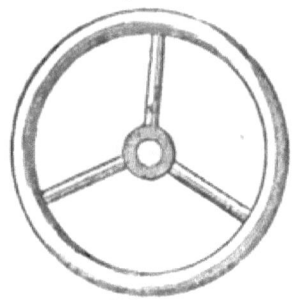

Fig. 1.

There are three factors upon which the time of the vibration of the balance depends:—

(1) The weight, or rather the mass of the balance.*

(2) The distance of its centre of gyration from the centre of motion, or, to speak roughly, the diameter of the balance. From these two factors the moment of inertia may be deduced.†

(3) The strength of the balance spring, or, more strictly, its power to resist change of form.

I append the usual formula for ascertaining the time of vibration of a balance, though it is difficult of application in actual practice:—

$$T = \pi \sqrt{\frac{AL}{M}}$$

A being the moment of inertia of the balance, M the moment of elasticity of the spring, L the length of the spring, and π 3·14159.

The Moment of Elasticity of a spring is its power of resistance. It varies directly as the modulus of elasticity of the material, and as the breadth and cube of the thickness of the spring when its section is rectangular. $Mo = E\frac{bt^3}{12}$ is a usual formula, E representing the modulus of elasticity, b the breadth, and t the thickness of the spring.

The moment of elasticity must not be confounded with the bending moment. The bending moment is a measure of the resistance a spring offers to bending, and of the amount of bending which has been produced, which varies directly as the angle wound through, and inversely as the length of the spring.

$M = \frac{Ebt^3 A}{L 12}$ is a formula for ascertaining the bending moment, E being the modulus of elasticity, b the breadth, t

---

* The mass of a body is the amount of matter contained in that body, and is the same irrespective of the distance of the body from the centre of the earth. But its weight, which is mass multiplied by gravity, varies in different latitudes.

† The centre of gyration is that point in a rotating body in which the whole of its energy may be concentrated. A circle drawn at seven-tenths of its radius on a circular rotating plate of uniform thickness would represent its centre of gyration. The moment of inertia or the controlling power of balances varies as their mass and as the square of the distance of their centre of gyration from their centre of motion. Although not strictly accurate, it is practically quite near enough in the comparison of plain balances to take their weight and the square of their diameter measured to the middle of the rim.

the thickness, and L the length of the spring, and A the angle through which it is wound.

This formula also determines the value of the force which has produced the bending, for if the forces are in equilibrium, the moment of the resisting force must be exactly equal to the moment of the bending force.

The Modulus of Elasticity is a constant, represented by E, which is used for ascertaining what proportion of its length material is strained when subjected to stress. If the body is stretched, the strain is a lengthening, and if it is compressed, a shortening of its original dimensions. In Young's formula, which is usually accepted, the stress in pounds per square inch of section, divided by E, gives the strain; E being the force in lbs. that would stretch a rod one square inch in section to twice its original length, supposing its elasticity to remain perfect all the time. Young gives 29,000,000 as E for steel, but Mr. Robert Gardner considers this too high for the average quality of steel used in balance springs, and places it at 23,000,000.

One end of the balance spring is fixed to a collet fitted friction tight on the balance staff, and the other to a stud attached to the balance cock or to the watch plate. The most ordinary form of balance spring is the volute or flat spiral, like Fig. 2. An overcoil or Bréguet spring is a volute with its outer end bent up above the plane of the body of the spring, and carried in a long curve towards the centre, near which it is fixed. (Fig. 3.) For marine chronometers helical springs, in which both ends curve inwards, are universally used. Either helical or Bréguet springs are as a rule applied to pocket chronometers, although a form of spring called "*duo in uno*," invented, I believe, by Mr. Hammersley, is sometimes preferred.

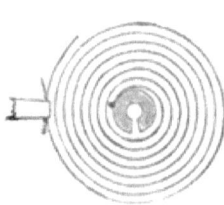
Fig. 2.—Ordinary balance spring.

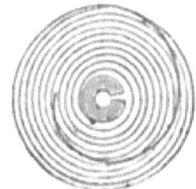
Fig. 3.—Flat spring with overcoil.

The bottom of this spring is in the form of a volute, from the outer coil of which the spring is continued in the form of a helix; the upper end is curved in towards the centre as in the ordinary helical spring.

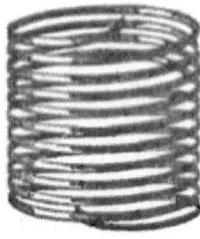 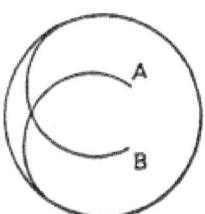

Elevation.   Plan.
Fig. 4.—Helical spring.

Among fancy shapes, which may be dismissed in a few words, are spherical springs, introduced by Frederic Houriet; they present no superiority, and having to be prepared on a solid spherical block are difficult to harden except by exposure to a high temperature. "Bird cage springs," having a helical body with top and bottom terminals formed into volutes, were another short lived conceit, with no advantage except the difficulty of making them.

The introduction of the balance spring which marks such an important epoch in the manufacture of watches is due to the investigations of Dr. Robert Hooke. He discovered that the potential energy of a spring is proportional to the angle through which it has been wound, and propounded the whole theory in the sentence, "*Ut tensio sic vis*," meaning that the force is proportionate to the tension. He proposed to patent his discovery in 1660, and to quote his own words, "Sir Robert Moray drew me up the form of a patent, the principal part whereof, viz., the description of the watch, is his own handwriting, which I have yet by me; the discouragement I met with in the progress of this affair made me desist for that time."

Derham describes the earliest of Hooke's essays in this direction as "a tender straight spring, one end whereof played backward and forward with the ballance." It is stated that several watches were made under Hooke's supervision at this period, and one of the first to which the balance spring was applied he is said to have presented to Dr. Wilkins, afterwards Bishop of Chester, about 1661.

It appears that Hooke then conceived it to be an advantage to have two balances coupled together, and had two double balance watches constructed. In the first, which had no balance spring, the escape wheel was placed in the centre of the movement with its teeth in a horizontal plane. There were two verges standing vertically on opposite sides of the wheel and connected with each other by means of toothed wheels of equal size; each verge had one pallet and carried a balance at its upper end, one balance overlapping the other.

In the second watch the verge escapement was arranged in the ordinary way, the balance being mounted on a verge with two pallets; on the verge was also a toothed wheel which engaged with another of the same size mounted on a stud, and the pipe of this wheel carried the second balance; the toothed wheels being of small size one balance was placed a little higher than the other and overlapped it. Each balance was controlled by a balance spring.

However, Hooke turned his attention to other matters, and in January, 1673, Huygens addressed a letter to Henry Oldenburg, secretary of the Royal Society, in which he described as his invention the application of a spring to control the balance in watches. This aroused the wrath of Hooke, who accused Oldenburg of having divulged the discovery in his correspondence with Huygens. Hooke enlisted the interest of Charles II., and in a lecture, entitled " Potentia Restitutiva," &c., said, " His Majesty was pleased to see the experiment that made out this theory tried at Whitehall, as also my spring watch."

A watch with two balances and springs, subsequently made for the King, was inscribed, " Robt. Hooke, inven : 1658. T. Tompion, fecit, 1675."

Hooke, Huygens and other investigators experimented with various materials to find the most satisfactory controller.

The engraving (Fig. 5) represents a watch of German origin from the collection of Mr. Evan Roberts. It has a day of the month ring, and is generally of the construction usual soon after the middle of the 17th century. But the peculiar feature of the movement lies in the application of a straight hog's bristle to regulate the balance. There is no

Fig. 5.—Hog's Bristle as a Balance Controller.

sign of any other spring having been attached, and the accessories of the bristle are quite in character with the rest of the work. There are two arms which embrace the bristle and practically determine its acting length, and by means of a screw these may be shifted to act over a considerable range.

Steel springs were however found to be the most suitable. The primitive straight ones would, of course, allow but a very small vibration of the balance, while the to and fro motion between pins where it made contact with the balance involved considerable friction. Of others curved somewhat to the shape of a pothook there are still examples, but eventually the more convenient and correct form was found to be a volute which at first had but one or two coils. The coils were increased to four or five as the advantage of a larger spring was understood, but the very long springs with which we are now familiar were not applied till the advent of the lever and other detached escapements which allowed the balance to have a larger arc of vibration.

To lengthen or shorten the acting length of the spring Tompion appears to have used the circular slide with an index from the first. This arrangement, which remained in favour for a long period, is shown in Fig. 6. Below, and attached to a silver disc, graduated and figured as a guide to regulation is a pinion which gears with teeth on the outer edge of the circular slide; from the inner edge projects an arm carrying two upright pins which embrace the spring. The projecting end of the pinion is square so that it could be turned by means of a watch key.

Fig. 6.—Tompion's Regulator.

Fig. 7.—Barrow's Regulator.

Mr. Albert Schloss has a clock-watch by Nathaniel Barrow, dating from about 1675, in which the outer end of the spring is continued in a straight line to the stud at the edge of the plate, and the regulation accomplished very much in the same way as the hog's bristle watch already delineated. Fig. 7 is a plan of this watch movement. The curved stud on the left is continued in a sort of zig-zag shape to hold one end of the regulating screw. The upper end of the nut points to an index engraved on the plate, and the lower extremity is notched to receive the spring.

The chief drawback to Tompion's regulator is that owing to the backlash or freedom between the teeth of the pinion and slide, a slight reversal of the index has no effect on the curb pins. The simple regulator now generally employed consists of a lever, fitting friction tight over a boss on the balance cock ; the shorter end of the lever carries the curb pins which embrace the balance spring, while the longer end through which it is moved serves also as an indicator of alterations in the position of the curb pins. This device was patented by Bosley in 1755.

There is one point about the stud used in those of Tompion's watches I have seen which might well be revived. The hole in the stud for the reception of the spring was square. The modern system of pinning by squeezing the flat side of a spring against the surface of a round hole is altogether unmechanical and must distort the spring.

The action of different forms of springs with opinions of various experts. — A very generally accepted rule is that the diameter of a steel balance spring for a watch should be half the diameter of the balance (rather under than over).

The dimensions of the spring, its form at the attachments, the position of the attachments with relation to each other, are all factors affecting its controlling power.

The length is important, especially in flat springs without overcoils By varying the thickness of the wire two flat springs may be produced, each of half the diameter of the balance, but of very unequal lengths, either of which would yield the same number of vibrations as long as the extent of the vibration remained constant ; yet if the spring is of an improper length, although it may bring the watch to time in one position, it will fail to keep the long and hort vibrations isochronous. Then, again, a good length of spring for a watch with a cylinder escapement vibrating barely a full turn would clearly be insufficient for a lever vibrating a turn and a half.

The great advantage of an overcoil spring is that it distends in action on each side of the centre, and the balance pivots are thereby relieved of the side pressure given with the ordinary flat spring. An overcoil spring, in common with the helical and all other forms in which the outer coil returns towards the centre, offers opportunities of obtaining isochronism by slightly varying the character of the curve described by the outer coil and thereby altering its power of resistance.

The position of the points of attachment of the inner and outer turns of a flat spring without overcoil in relation to each other has an effect on the long and short vibrations quite apart from its length. For instance, a different performance may be obtained with two springs of precisely the same length and character in other respects, but pinned in so that one has exactly complete turns, and the other a little under or a little over complete turns. This property, which is more marked in short than in long springs, is depended upon by many for obtaining isochronism. A short spring as a rule requires to be pinned in short of complete turns, and a long one beyond the complete turns. In duplex and other watches with frictional escapements, small arcs of vibration and short springs, it will be found that the spring requires to be pinned in nearly half a turn short of complete turns. In watches the point of attachment to the collet with relation to the pendant has also to be considered with plain flat springs and with flat spring springs having an outer terminal curve. It is not easy to see why a helical or other spring with two theoretically correct terminal curves should be affected by varying the positions of the points of attachment with relation to each other, yet marine chronometer springs are found to isochronize better and act truer when pinned in at about a quarter of a turn short of complete turns.

If a spring is too long the short vibrations will be slow. It may be taken as a very good rule that a steel balance spring should be half the diameter of the balance, and have twelve turns if it is a flat spring or eighteen turns if a Bréguet for a lever watch of ordinary size. Small watches are more difficult to time than those of a reasonable size, and are generally slow in the short arcs, being relatively more affected by the retarding action of the escapement. For small lever watches a spring shorter by a turn or two will therefore be desirable. These lengths, it will be understood, only apply where the work is good; with coarse work a shorter spring is usually required in order to get the short arcs fast enough. Springs for cylinder watches should have from eight to twelve turns.

Watch springs of thick and narrow wire are apt to cockle with large vibrations, while springs of wide and thin wire keep their shape and are more rigid. It is of even greater importance that the springs of marine chronometers subjected to the tremor of steamships should be of wide and thin wire.

Helical springs for ordinary two-day marine chronometers are made from ·4 to ·54 of an inch in diameter, and about a quarter of a turn short of either eleven, twelve, or thirteen turns.

It is remarkable that while in watches the difficulty is generally to get the short arcs sufficiently fast, precisely the reverse is the case with the marine chronometer, in which the trouble is usually to get the short arcs slow enough. The escapement is not entirely responsible for the difference, because pocket chronometers follow the same rule as watches with lever escapements.

The relative slowness of the long arcs in marine chronometers is much greater if the rim of the balance is narrow and thin, and the weights large, than if a wider and stouter rim with proportionately smaller weights is used; for in the former case the greater enlargement of the rim in the long arcs from centrifugal tendency will be more marked. Mr. Kullberg, by substituting a chronometer balance of ordinary proportions for an uncut one, has demonstrated that the effect of centrifugal tendency by increasing the arc of vibration from three-quarters of a turn to a turn and a quarter amounts to twelve or fourteen seconds in twenty-four hours. Mr. B. Dennison some years ago advocated a method of making the long arcs slower in watches having compensation balances, by drilling a small hole in each half of the balance rim, close to the arm, and broaching these holes out as much as should be found necessary to produce isochronism. The larger the holes, the more the balance would expand from centrifugal tendency, which, of course, would have more effect in the long than in the short arcs.

The size of the pivots in proportion to the size of the balance is partially the cause, for in very small watches, where of course the pivots are relatively large, the slowness of the hanging position is proverbial, and a shorter spring by a turn or two has often to be substituted. Very quick trains should be avoided on this account. Watches have occasionally been made with 19,800 vibrations, carrying of course corresponding light balances, and the great trouble has always been to get them fast enough in the short arcs.

A balance staff pivot slightly too large for the hole is occasionally the cause of undue slowness in the hanging position of a watch; a very small reduction in the diameter will in such cases reduce friction and quicken the short arcs.

Mr. Robert Gardner insists that the relative proportional inertia of the fourth wheels in marine chronometers and watches accounts for much of the difference observed in the long and short arcs, and there is no doubt that watches with small fourth and escape wheels are comparatively easy to time. J. F. Cole suggested a stronger mainspring to quicken the short arcs in going-barrel watches.

With single beat escapements, such as the chronometer and duplex, the short arcs are quickened if the drop of the escape wheel tooth on the pallet is decreased; if the drop is increased the long arcs are quickened. The theory of this appears to be that with more drop in the long arcs (when the balance is travelling faster) the pallet gets further away before being overtaken by the wheel than it does in the short ones, and therefore the amount of impulse given before the line of centres is proportionately less; and Mr. Robert Gardner asserts that the same effect may be produced by putting the piece out of beat, *e.g.* if it were desired to quicken the short arcs, the balance spring collet would be shifted so as to bring the pallet more away from the unlocking when the balance spring was quiescent; but it must not be forgotten that if the piece is out of beat it is more likely to set.

Fig. 8 is a careful reproduction of the curves of a marine chronometer spring. *a* is the lower coil. The two short lines crossing the spring denote the face of the collet and the face of the stud respectively, and the dotted lines the direction of the ends of the spring, which form nearly a right angle. Occasionally, if the short arcs are fast, the upper turn is slightly bent just as it enters the stud so as to throw the end outward. The pin is then placed on the inside.

Fig. 9 is a lower coil, and Fig. 10 the upper coil of a pocket chronometer spring. It will be observed that the sweep is longer than in the marine chronometer, and that the lower curve is carried rather farther back into the spring than the upper. The spring makes complete turns, but the upper turn just as it enters the stud *has a slight sharp bend*

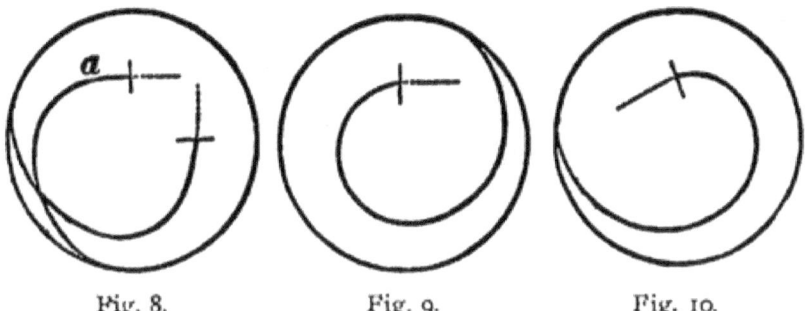

Fig. 8. Fig. 9. Fig. 10.

throwing the end inwards. This little bend is of the utmost importance, for it has the effect of quickening the short arcs. Pocket chronometer springs are made half the diameter of the balance, and from seven to ten turns.

Fig. 10 will also serve very well to represent the overcoil of a Bréguet spring, although the overcoil is sometimes carried much farther into the spring. In some springs it is not unusual to see the overcoil more than a complete turn in extent. Bréguet springs are now often used for pocket chronometers, instead of the helical form. Just as good a result can be got with the Bréguet as with the helical, and the latter takes up height, and in consequence is often made too short. There is, though, one advantage in using helical springs for pocket chronometers. The escapement may be banked through the spring, and this is done more readily in the helical form. Mr. Kullberg's method is to place two nearly upright pins on the balance arm, so close to the spring as to prevent it expanding more than is required for a sufficient vibration. These pins should be slightly inclined to the centre so as to touch the top of the spring first, and thereby stop the balance more gradually.

Two pairs of pliers with curved noses lined with brass are generally used for forming the overcoil of marine chronometer springs. The coil of the spring where the overcoil is to start is grasped by one pair curved exactly to correspond with the spring, and the other pair used to bend the overcoil. **The** operation **looks** easy enough, but it really requires great skill to get at once an overcoil of the desired shape. The overcoils of watch springs are turned with curved nosed steel tweezers.

Acceleration.—It is noticed that **new** chronometers and watches, instead of steadily gaining **or** losing a certain number of seconds each day, go faster day after day. There **is** no certainty as to the amount or ratio of this acceleration, nor as **to the** period which must elapse before the rate becomes steady, **but** an increase of **a** second a month for a twelvemonth may be taken **as** the average extent in marine chronometers.

It is pretty generally agreed among chronometer-makers that the cause of acceleration **is** seated in the balance spring, though some assert that centrifugal action slightly enlarges the balance if the **arc of** vibration is large, as it **would** be **when** the oil is fresh, **and** that as the vibration **falls** off **centrifugal** action is lessened **and** acceleration ensues from **the** smaller diameter of the balance. Though thin balances **do** undoubtedly increase slightly **in size in** the long vibrations from centrifugal action, this **theory is** disposed of by the fact that old chronometers do **not** accelerate after re-oiling. **Others** aver that the unnatural connection of the metals composing the compensation balance is responsible for the mischief, and that after being subjected to heat the balance hardly returns to its original **position** again. If true, **this** may be **a reason for exposing new chronometers** before they are rated **to a somewhat higher temperature than they** are likely to meet **with** in use, **as is the practice of some makers;** but **then** chronometers **accelerate on their rates when they are kept in a** constant **temperature, and also if a new spring is put to** an old balance, or **even if a plain uncut balance is used.**

It is noticed that when the overcoil of a balance spring has been much bent or "manipulated" in timing, the acceleration is almost sure to be excessive. This is just what might be expected, for a spring unduly bent so as to be weakened but not absolutely crippled, recovers in time some of its lost elasticity. But however carefully a spring is bent the acceleration is not entirely got rid of, even though the spring is heated to redness and again hardened after its form is complete. There is little doubt that the tendency of springs is to increase slightly in strength for some time after they are subjected to continuous action, just as bells are found to alter a little in tone after use. As a proof that the acceleration is due to the bending of the overcoil, Mr. Hammersley asserts that, if the spring of an old chronometer is distorted and then restored to its original form, the chronometer will accelerate as though it were new. Helical springs of small diameter have been advocated by some as a means of lessening acceleration, on the ground that the curves are less liable to distortion in action than when the springs are larger. Mr. T. Hewitt tells me it was noticed that if the balance springs were tempered on a larger tube than the one they were hardened on, as was formerly the practice of some makers, the acceleration was sure to be excessive. Springs elongate in hardening, and it has been suggested that they afterwards gradually shorten to their original length and so cause acceleration, but there does not seem to be much warrant for this assumption. Unhardened springs do not accelerate, but then they rapidly lose their strength, and are therefore not used. Flat springs do not accelerate so much as springs with overcoils. Palladium springs accelerate very much less than hardened steel springs.

Having summarized the various opinions of practical experts, let us see how far they are founded on scientific bases.

CHAPTER II.

Theoretically correct terminal curves.—Effect of disturbing influences.—Of late years science has lifted the veil which formerly obscured the reasons for various operations in connection with springing which were accepted as the correct procedure solely from the result obtained. In many instances these empirical dicta were founded on sound bases, but there is always the danger in accepting rules given as "practical experience" that the result observed has been merely the accidental presence of unsuspected factors.

M. Phillips, a distinguished French mathematician, investigated the laws governing the action of the balance spring and in "*Mémoire sur le spiral réglant,*" &c., which was printed in 1861, published the result of his researches; giving a number of examples of theoretically correct terminal curves.

Jules Grossmann, professor at the Locle School of Horology, in continuation of Phillips' labours, examined various factors connected with the balance spring as in operation in a timekeeper, and brought, as it were, the subject into the domain of the watchmaker.

L. Lossier reduced to greater simplicity some of Jules Grossmann's results, and produced a most excellent manual of the Theory of Timing. His work, ably translated from the "*Journal Suisse d'Horlogerie,*" by Mr. George Walker and Mr. W. N. Barber, appeared a short time ago in the "*Horological Journal.*"

Several other writers have contributed to the sources of theoretical and practical information on the subject as will appear from acknowledgments made in due course.

The first requirement for a spring to be isochronous is that the centre of gravity of the spring shall be coincident with the centre of gravity of the balance.

In a cylindrical spring, furnished with terminal curves we have to consider two portions; first, the whole of the coils forming a certain number of complete turns commencing and ending at the same point C (Fig. 11), and whose centre of gravity falls exactly on the centre of the figure O, then the part composed of the two curves $C A$ and $C' A'$, and the added part of the spring $C D C'$ of any length whatever.

In order that the whole spring may have its centre of gravity on O, it is only necessary to form the curves in such a manner that the three portions of the spring $A C, C C'$, and $C'A'$, are in equilibrium, and have a common centre of gravity in O.

To do this we shall have to find the positions of the centres of gravity of G and G' which satisfy these conditions.

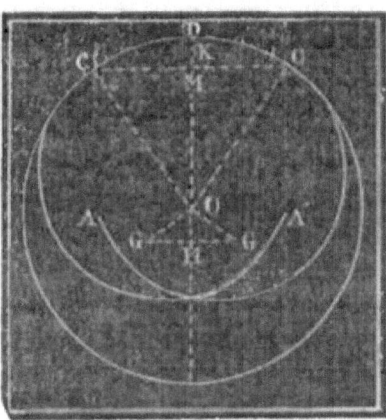

Fig. 11.

I do not propose to follow the demonstration of this, which may be obtained from various text books on Graphical Statics, but it leads to Phillips' formula:—

$$O\,G = \frac{R^2}{l}$$

This may be stated as but involving the following two conditions:—

1st. That the centre of gravity of the curve be on the line of $O\,G$, making a right angle with $O\,C$.

2nd. That the distance of the centre of gravity of the curve from the centre of the spring be equal to the square of the radius of the spring divided by the length of the curve.

Phillips demonstrated that when the centre of gravity of the spring has been thus established by construction, on the axis of the balance, it will remain there whatever may be the deformation of the spring, so that this latter, either in contracting or expanding, always remains concentric with itself.

He also gave a method of graphically constructing the theoretic curves, and Lossier has devised a practical rule of correction, which may be with advantage adopted, and which I venture to reproduce.

The two terminal curves of the same spring need not be identical in form and dimensions. If, in fact, one of the curves is, for example, longer than the other, it will be necessary, in order to satisfy the formula of Phillips, that its centre of gravity be nearer to the centre of the spring in order that its moment with relation to the centre may remain the same.

Let us suppose that one of the curves is twice as long as the other, so that its length will be $2\,l$; in order that it may comply with the conditions of Phillips' formula, since R does not change, the distance of its centre of gravity from the centre of the spring should be but one half that of the other curve, that is $\frac{O\,G}{2}$. The moment of this curve will then be $2\,l \times \frac{O\,G}{2}$, or $l \times O\,G$, which is the same as the moment of the other curve, therefore the equilibrium of the whole spring will not be modified.

We see, consequently, that we may change the length, the form, and the position of the curves as we like, provided that the condition holds for each of them, that $O\,G$ be equal to the square of the radius, divided by the length of the curve.

The theory of Phillips adapts itself equally well to the flat spring whether it be for the exterior curve, which does not sensibly differ in form from that of a cylindrical spring, or for the interior curve. This latter ought to be considered as composed of that part of the spring joining the largest and the smallest coils and of a connecting curve with the collet.

It results from this that the small connecting curve cannot of itself alone comply with Phillips' conditions, since it has also to agree with an archimedian spiral. However it need not differ much from it, but the difficulty of its exact reproduction from a drawing increases considerably with its smallness. Fortunately as we shall see this inner curve may in most instances be dispensed with, but if it is desired,

Lossier suggests forming a curve something approaching the theoretical shape and then correcting it as the action of the spring when the watch is going may indicate.

It should be noticed that when a spring is furnished with Phillips' curves, not only is its centre of gravity not displaced, but the spring itself expanding and contracting in a uniform manner on all sides, does not exert any pressure or any effect of torsion on the axis of the balance.

It will be understood that no particular form of curve is prescribed by Phillips. So long as the curve conforms to the conditions recited the character of the curve is unimportant. Appended are four of several examples given by Lossier. In all a spring theoretically imperfect distending towards N, is represented by dotted lines and a suggested alteration of the outer terminal renders it theoretically perfect.

These springs are all right handed, that is the volute developes, starting from the left of the centre O, upwards to the right. And with such springs the isochronal correction is made by drawing that part of the curve to the right of the centre line $M\ O\ N$, towards the centre, or the part

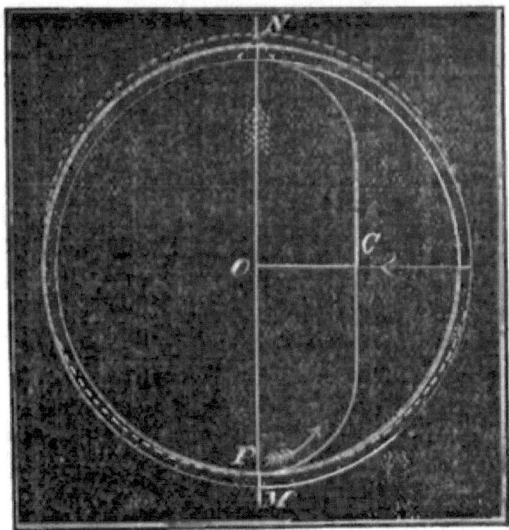

Fig. 12.

to the left of the centre away from the centre, or by a

combination of these operations. With left handed springs the correction would of course be in the reverse direction.

In Fig. 12 an overcoil is formed of two quarter circles having a radius equal to half the radius of the spring $O\ P$ joined by a straight line C. It will be observed that in this case the termination of the overcoil is not carried in towards the centre so that the stud remains in the same position as for a flat spring. Mr. Kullberg adopted a somewhat similar overcoil but with a short outer concentric termination for use with curb pins and he spoke well of watches so fitted.

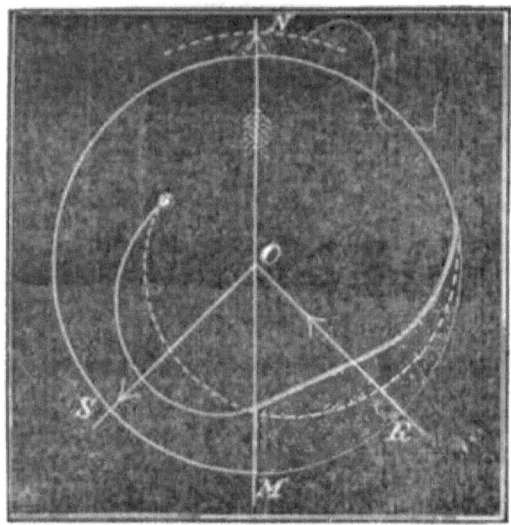

Fig. 13.

In Fig. 13 the original curve which was not isochronal was formed of half-a-circle as dotted. The full lines show the alteration to an isochronal overcoil without disturbing the point of attachment to the stud. The part to the right of the centre line has been carried towards O along the line $R\ O$, while the part to the left has been drawn out from O towards S. In this compound correction the lines $O\ S$ and $R\ O$ forming the centres of the respective corrections should form a right angle.

In Fig. 14 the curve has been carried in from the right, and in Fig. 15 carried out to the left of the centre.

Lossier gives a curve, easy to construct, fulfilling the

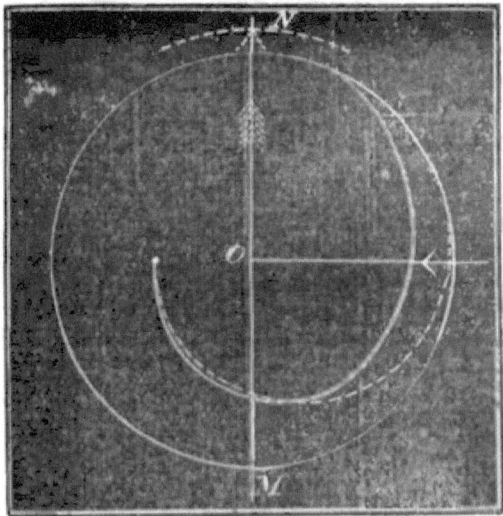

Fig. 14.

condition necessitated by the regulator, and which may serve as a basis for the exterior curve of a flat or a cylindrical spring, subject to slight modification afterwards, according to the performance of the watch.

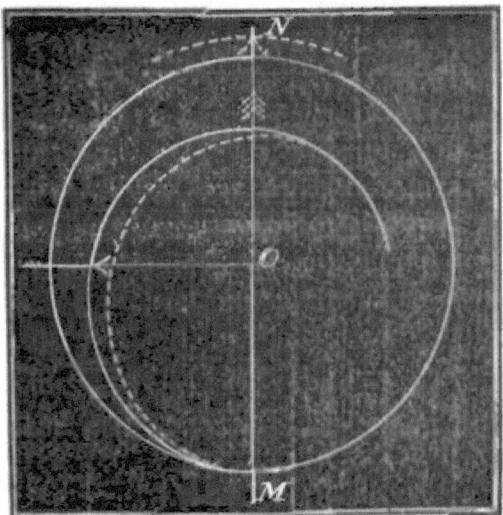

Fig. 15.

From the centre of the spring (Fig. 16), trace an arc of a circle with a radius of ·67 R (R being the radius of the exterior coil), the arc ought to comprise beyond the curb pins an angle of of 83°, and connect it with the exterior coil by a semi-circle of a radius

$$\frac{1\cdot 67\,R}{2} = \cdot 835\,R.$$

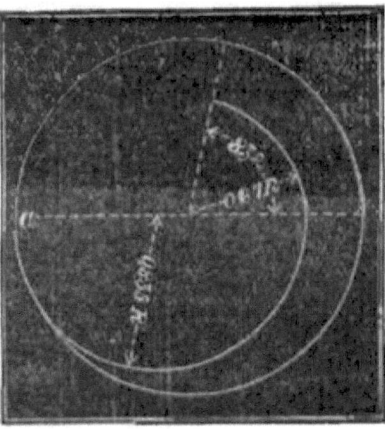

Fig. 16.

This curve may serve for the internal curve of either a cylindrical or flat spring. For this latter, a space of about four coils is left between the collet and the interior coil, connection being made by a coil constructed like that in Fig. 16, taking for R the radius of the internal coil. The part bc (Fig. 17), of the coil being concentric with the collet would touch this latter, which would completely destroy the isochronism. To avoid this inconvenience, the extremity is curved in such a manner that there remains a space, equal to the distance between two coils, between the collet and the beginning of the curve.

These curves only give a concentric development when they are reproduced with exactitude; but as it is extremely difficult to obtain the desired precision in the execution, and besides there may arise circumstances obliging one to deviate from the rigorously theoretical form, it is practically more simple to reproduce these curves as near as possible by the eye, and then correct them according to the action of the spring.

Method of Forming and Testing Terminal Curves.—Messrs. George Walker and W. N. Barber have devised a way of forming the curve, together with a sort of scale for testing it. According to Phillips the conditions to be satisfied by a terminal curve are :—

1st. That the centre of gravity of the curve fall on the line OB making a right angle with ON (N being the starting point of the curve).

2nd. That the distance from the centre of gravity of the curve to the centre of the spring be equal to the radius of the spring squared, divided by the length l of the curve.

$$\text{Then } OG = \frac{ON^2}{l}$$

Take the radius of the spring ON as a constant, and l of any convenient length determined by experience; from these data calculate the distance of the point G from the centre of the spring. From the centre O draw a circle having a radius equal to the distance of the inner side of stud hole from the axis of the balance (the same scale being used as for the rest of the construction.) From centre G draw a circle of such a radius as will include the point N; cut this circle out of uniformly thick material; this circle will then balance on the point G.

Take a rod of uniform section of the pre-determined

length of the curve and place one end on the point N, the other end on the circle drawn for the stud hole; this rod may then be bent to any form so that the whole system balances on the point G, when the conditions for isochronism will be satisfied.

By means of Messrs. Walker & Barber's Tester the moment of a terminal curve with reference to the axis of the spring may be weighed against a mass so placed that its moment is equal and opposite to that of the terminal curve.

A circular plate is accurately balanced on a centre which has a small vertical movement in a tripod stand fitted with levelling device, a stepped annular ring embraces the balanced plate, so that when the centre is lowered the plate

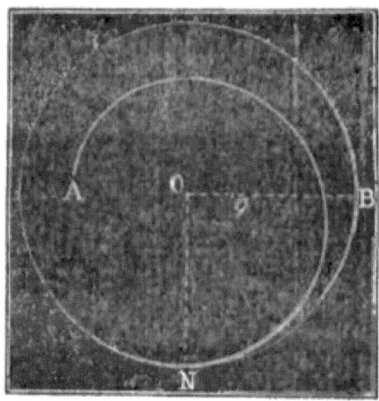

Fig. 18.

rests on the step of the annular ring, thus allowing manipulation of the terminal curve without danger of damage to the centre. The coincidence of the top surface of the ring and the balanced plate shows when the whole is in equilibrium.

The terminal curve is arranged on the surface of the balanced plate, the centre of the plate representing the axis of the spring. Having determined the two circles on which the ends of the curve will lie, and calculated the position of the centre of gravity of the curve as before, the moment of the curve may be found by multiplying the weight of the curve by the distance of its centre of gravity from the axis. By placing a mass, say equal to the weight of the curve, on the line *BO* produced, and at the same distance from *O* as the centre of gravity of the curve, the theoretical conditions will be satisfied when the whole system is in equilibrium.

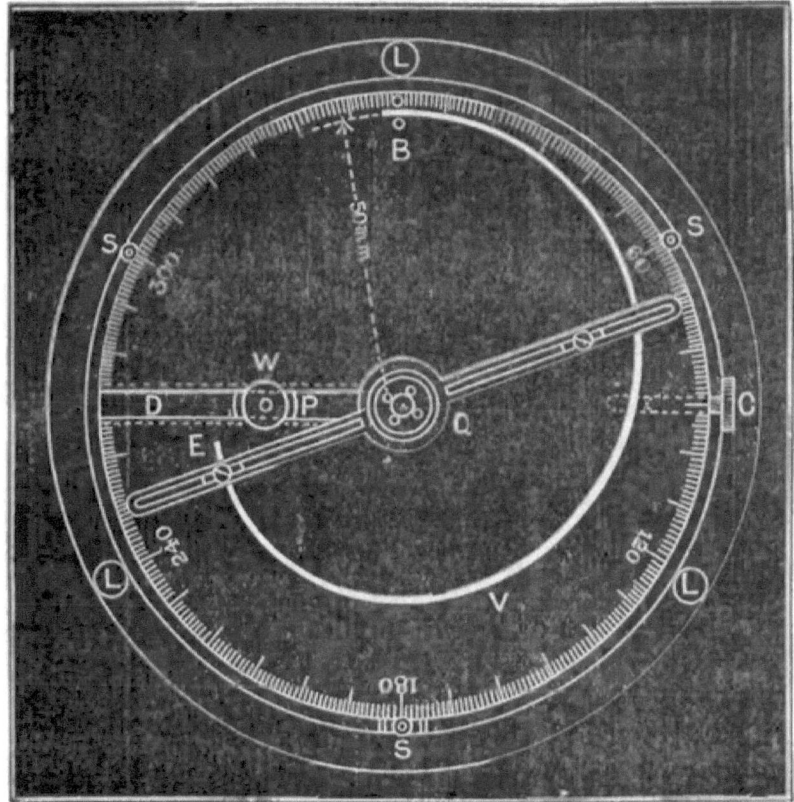

Fig. 18A, Plan of Walker and Barber's Tester.

PLAN (the same objects are represented by the same letters in plan and elevation):—

    B V E=Terminal curve.

        D=Dovetail slide.

        P=Dovetail slip with vertical pin, to take weights, which may be adjusted at any distance from the centre o. the plate.

        C=Counterweight, to secure adjustment of the plate when the slip P is moved to different positions.

ELEVATION R M=Stand with levelling screws.

        L, L, L.=Levelling screws.

            N.=Balanced plate on which the terminal curve is adjusted.

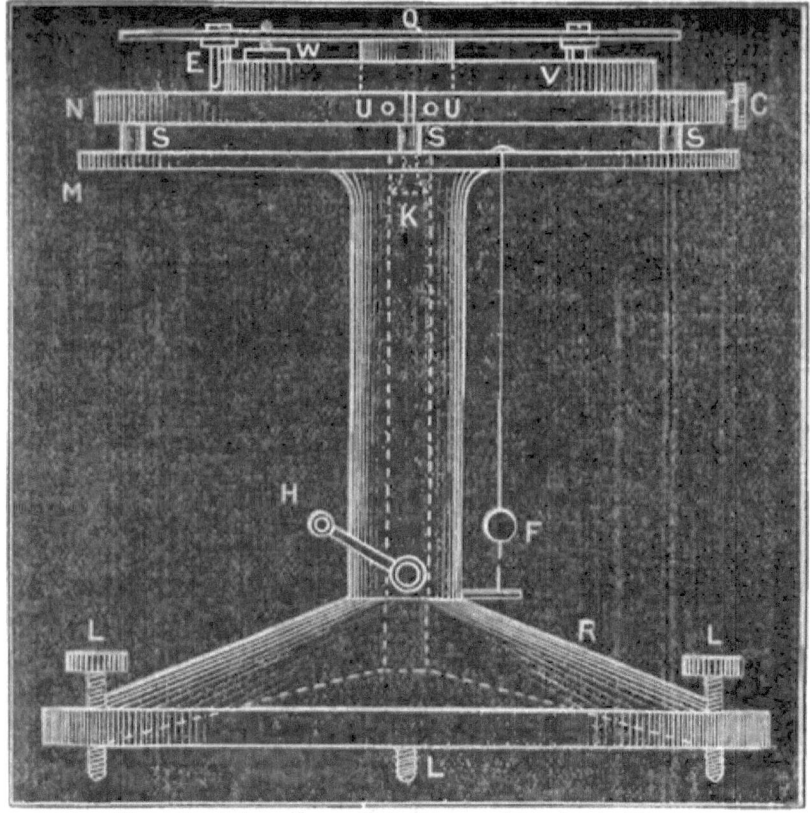

Fig. 18B, Elevation of Walker and Barber's Tester.

S. S. S. = Studs, screwed into the stand, on which the balanced plate rests while the terminal curve is being manipulated.

E V. = Terminal curve resting on balanced plate.

W = Weight, to secure equilibrium when the curve is correctly formed.

Q = Double-ended bar, with boss, which is snapped into recess in balanced plate, and carries slides with pins to keep the end of the terminal curve at right distance from the axis.

C = Counterweight.

F = Plumb-line, for levelling apparatus.

K = Pump centre, shown dotted.

H = Handle of eccentric for raising and lowering pump centre.

U.U. = Pins, to prevent circular motion of balanced plate.

To use the Tester.—Let us suppose that the terminal curve is resting on the balanced plate, and arranged to fulfil the theoretical conditions. Its C of G will be at a point G at a certain distance from the centre O and on a line at right angles to a line joining the centre O to the beginning of the curve, measured round the curve. If the plate be free to move it will turn round the axis O in such a direction that the line G O moves in a vertical plane with a moment G O + the weight of the curve.

If in the same line, on the opposite side of the axis O, a weight W be placed such that its moment is equal to that of the terminal curve the plate will be in equilibrium. Now if the terminal curve be changed in shape the equilibrium will be destroyed, and the curve will no longer fulfil the theoretical conditions.

To design any terminal curve, taking 100 mm. as a standard diameter, it is only necessary to know the length and weight of curve from which the requisite moment to produce equilibrium may be found, and at the same time fulfil the theoretical conditions.

The terminal curve should be made of some easily bent and heavy metal to facilitate manipulation.

A spring with theoretically perfect terminal curves would be correct if applied to a balance also perfect, and not subjected to external influences. But such conditions do not, of course, exist in watch work, and, therefore, the consideration of all the factors with which the practical horologist has to contend is the next step.

AN EXTERNAL FORCE APPLIED TO THE BALANCE AT THE DEAD POINT, THAT IS WHEN THE SPRING IS QUIESCENT, HAS NO EFFECT ON THE TIME OF OSCILLATION, AND THE EFFECT OF A FORCE IS MUCH ACCENTUATED AS THE DISTANCE BETWEEN THE DEAD POINT AND THE POINT OF APPLICATION INCREASES.

AS A RULE THE DISTURBING EFFECT OF A GIVEN EXTERNAL FORCE WILL VARY ACCORDING TO THE AMPLITUDE OF THE OSCILLATIONS, AND WILL BE MUCH MORE MARKED AS THE AMPLITUDE DIMINISHES.

AN IMPULSE DELIVERED BEFORE THE BALANCE REACHES ITS DEAD POINT QUICKENS THE TIME OF VIBRATION.

AN IMPULSE DELIVERED AFTER THE BALANCE HAS PASSED ITS DEAD POINT EXTENDS THE TIME OF VIBRATION.

SIMILARLY, ANY ABSTRACTION OF FORCE IF TAKING PLACE BEFORE THE DEAD POINT IS REACHED EXTENDS THE TIME OF VIBRATION, WHILE ABSTRACTION OF FORCE AFTER THE DEAD POINT IS PASSED QUICKENS THE VIBRATION.

IN THE CYLINDER ESCAPEMENT AND IN THE LEVER ESCAPEMENT ALSO THE IMPULSE IS NOT EQUALLY DIVIDED ON BOTH SIDES OF THE DEAD POINT, THEREFORE, SO FAR AS THE INFLUENCE OF THE ESCAPEMENT GOES ALL CYLINDER AND LEVER WATCHES MUST LOSE IN THE SHORT ARCS.

With these data let us examine what takes place in a cylinder escapement, taking the angle of vibration as $\frac{2}{3}$ of a turn. Let the angle of lift be 40°, of which 5° are for locking, leaving 35° for impulse. The balance, starting at 120° from the dead point, returns to within 15° of this point without being subject to any other influence than friction. From 15° to 0° it will receive impulse in the same direction as its movement, and consequently there will be an acceleration. After the dead point is passed, the impulse will continue for 20°; the first 15° will produce a retardation equal to the preceding acceleration, thus leaving 5° during which the retardation produced will not be compensated.

The final effect of the lift will then be to increase the time of the oscillations, because the angle of repose necessitates a longer impulse on one side of the dead point than on the other. The watch will thus lose more in the short than in the long arcs, more in the hanging than in the lying position.

In the lever escapement the minimum amount of locking is 1°, measured from the axis of the pallets; and we may take 2° as the maximum admissible. To determine the same angle of locking, considered with relation to the movement of the balance staff, we must replace the angle measured at the pallet staff by the corresponding angle moved through by the balance. Multiplying the former by the ratio between the length of the lines representing the lever and roller, according as this ratio is 4½, 4, 3½, 3, the angle moved by the balance, the pallets having 1° of locking, will be 4½, 4, 3½, or 3 degrees; and for 2° of locking it will be 9, 8, 7, or 6 degrees.

If the locking at the pallets be 2°, and the ratio between the length of the lever and that of the roller is 4½, the effect of the locking commences at 22½° from the dead point (for a lift of 10° at the pallets), and ceases $22\frac{1}{2}° - 9° = 13\frac{1}{2}°$ from the dead point. If this ratio be 3 : 1, the effect commences about 15° and ends at $15° - 6° = 9°$ from the dead point. As the effect of a force acting on a balance depends on the angular distance of the point of application from the dead point, the effect of the angle of repose on the timing will be less in the second case than in the first.

The advantage of escapements, giving a small arc of connection with the balance, is apparent. An ideal escapement would be one delivering the whole of the impulse by a blow at the dead point. No such escapement exists, and in our endeavour to approach this ideal the motion of the pallets and arc of connection with the balance are sometimes too much diminished, leaving insufficient force to maintain the vibrations.

There is a limit to variations in the proportion of the lever escapement, though it is not precise. Small locking angles are desirable, but the locking must be safe for all the teeth of the wheel. For the same balance arc Swiss escapements have usually a lesser lifting angle for the pallets than English, and a longer lever. With too great a lifting angle the tooth is likely to set on the impulse planes, and with too little to set on the locking.

There is another variable disturbing factor connected with the lever escapement, and that is the drop on to the impulse pallet after the unlocking, which affords an exception to the general proposition that all disturbing influences have less effect in the long than in the short vibrations. Even with the lightest escape wheel there must be an interval between the unlocking of the tooth and its contact with the impulse plane, and owing to the draw on the locking face the wheel receives a recoil at each unlocking which lengthens this interval. In the long vibrations the unlocking is more energetic and the recoil greater, and, as the balance is moving quicker, the pallet will evidently get farther away before the tooth reaches it than it would in the short vibrations. The loss of impulse before the dead point from this cause will therefore be greater in the long than in the short vibrations, and the effect will be much more marked if the escape wheel is large and heavy, because then its inertia will render its action more sluggish, and the point where it impinges on the impulse pallet will be farther along the plane than would be the case with a smaller and lighter wheel.

With the chronometer and other single beat escapements the variation of the time of oscillation of the balance is not necessarily so pronounced. Lossier suggests shifting the dead point in the chronometer escapement, so that when the spring is at rest the impulse pallet is nearer to the exit tooth. The impulse may be equalized on each side of the dead point in this way, and the short arcs quickened thereby, but a chronometer out of beat is more likely to set.

Point for attachment of the spring to the collet.—The position of the point where the spring is attached to the collet WITH RELATION TO THE PENDANT OF THE WATCH is of prime importance, except in helical or other springs provided with two isochronal curves. The centre of gravity of a flat or other form of balance spring without an internal terminal curve and pinned to the collet in the usual way is continually shifted during the action of the spring, and it will be evident that its poise, and practically the poise of the balance, will be affected during the going of the watch. If the spring were affected equally in its coiling and uncoiling this displacement might be ignored, but in Lossier's treatise is a demonstration by Jules Grossmann which shows that a spring distends more from its normal radius in uncoiling than it contracts in coiling.

Advantage may be taken of this fact to neutralize in many instances positional errors, and to quicken the short arcs.

The exact effect of varying the point of attachment cannot be invariably predicated, but generally with a spring developing to the right as in Fig. 19 the result was found to be as follows:—

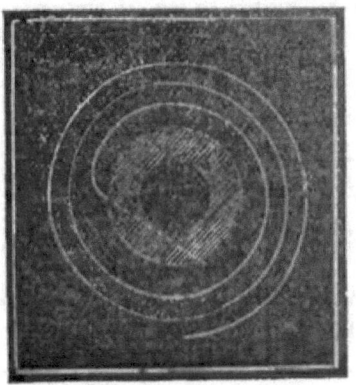

Fig. 19.

Position of the point of attachment.	Pendant			
	left,	right,	up.	down.
Left ..	o	o	gain	loss
Right ..	o	o	loss	gain
Up ..	loss	gain	o	o
Down ..	gain	loss	o	o

Palladium springs, being heavier, would show a greater result from the shifting of the centre of gravity than steel springs.

THE PRACTICAL RESULT IS THAT A WATCH WITH A BALANCE SPRING DEVELOPING UPWARDS TO THE RIGHT, AS IN FIG. 19, HANGING (OR PENDANT UP), AND LOOKING AT THE BACK THE POINT OF ATTACHMENT SHOULD BE TO THE LEFT OF THE BALANCE CENTRE AND ON A HORIZONTAL LINE WITH IT.

To facilitate obtaining this position Jules Grossmann suggested drawing on a sheet of paper two lines at right angles (A, B, and C, D, Fig. 20), and placing the movement so that the centre of the balance coincides with the point of intersection, and that a line, E, F, parallel to A, B, passes through the centre of the dial. The line G touches the front of the stud and may be continued to the centre of the balance O, as in Fig. 21. T is the point for the attachment to the collet, and the point for the stud at P can be marked on the spring.

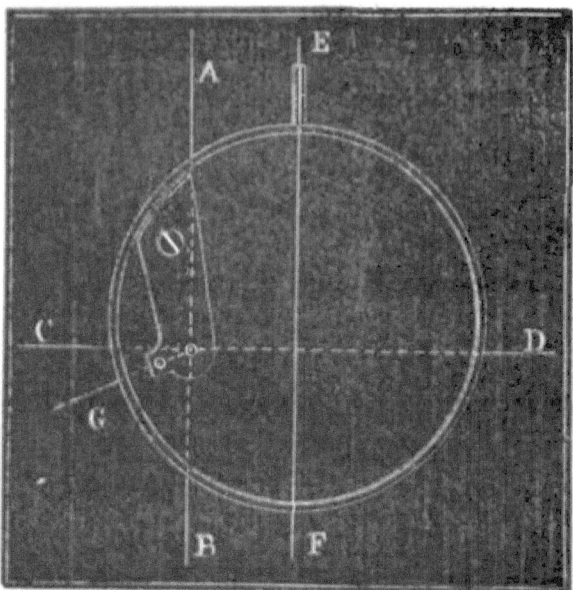

Fig. 20.

Occasionally it may happen from some constructional defect that with the spring so pinned in there is a loss on turning the watch pendant left. In such a case the plan would be to cut a piece say seven-eights of a turn out of the

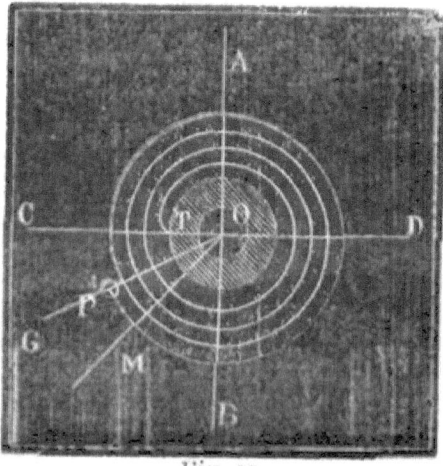

Fig. 21.

eye of the spring so as to bring the point of attachment towards M.

There appears to be but little doubt that so far as flat springs with an exterior overcoil are concerned the old and contradictory data as to pinning in springs at complete coils, or short or over complete coils really amounted to nothing more than getting the point of attachment to the collet into the most favourable position. Cutting and repinning the spring, as here recommended, in order to shift the point of attachment could be avoided if the stud were capable of being moved round concentrically with the balance. Messrs. Walker and Barber and Messrs. P. & A. Guye have devised adjustable studs of this description, but in the majority of watches cutting and repinning the spring must be resorted to.

For a left-handed spring or one developing from the centre upwards to the left, as in Fig. 22, the result was as follows:—

Position of the point of attachment.		Pendant			
		left,	right,	up,	down.
Left	..	o	o	loss	gain
Right	..	o	o	gain	less
Up	..	gain	loss	o	o
Down	..	loss	gain	o	o

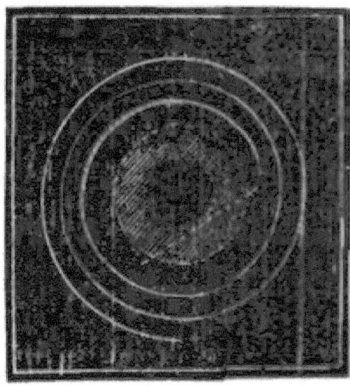

Fig. 22.

With a flat spring having no overcoil there is another factor which should not be passed unnoticed. Here the spring in coiling and uncoiling is subjected to a pushing from and drawing towards the stud which is tantamount to an additional winding or unwinding of the spring.

If the spring consists of complete turns the stud and collet attachments are both on the same side of the centre, and then there is no tendency to twist the collet when the spring starts from its point of rest as there would be if the spring were, say, a quarter of a turn short, or a quarter of a turn over complete coils. In action the relative position of the attachments is continually altering, and the effect in any particular case is difficult to calculate.

The general conclusion shown by Lossier is in favour of pinning in at a quarter of a turn over complete coils as giving the best result in vibrations of ordinary amplitude; a quarter of a turn short showing a more even performance than either complete turns or half turns.

With an adjustable stud a flat spring without overcoil may be applied so as to bring the collet attachment in a line with the centre of the balance in accordance with the prescription given for obtaining the best result in positions, and also to make a quarter of a turn over complete coils. But with the position of the stud fixed one of these desiderata may have to be sacrificed.

CHAPTER III.

Compensation for varying Temperature.—John Harrison, in his efforts to obtain the Government reward of £20,000 offered for a timekeeper sufficiently accurate to determine the longitude, was the first to apply a corrective to the balance spring for the temperature variations.

One of the chief features of his celebrated watch, now preserved at the Royal Observatory, Greenwich, is a bimetallic arm fixed at one end, and carrying at its free ends two pins, to embrace the balance spring near its outer point of attachment. This is his description of the device: " The thermometer kirb is composed of two thin plates of brass and steel riveted together in several places, which, by the greater expansion of brass than steel by heat, and contraction by cold, becomes convex on the brass side in hot weather, and convex on the steel side in cold weather; whence, one end being fixed, the other end obtains a motion corresponding with the changes of heat and cold, and the two pins at this end, between which the balance spring passes, and which it touches alternately as the spring bends and unbends itself, will shorten or lengthen the spring."

Harrison at first provided additional curb pins for mean time adjustment, but had to abandon them; for it is clear, if they were placed behind the pins on the compensation curb, they would not act, and, if placed in front, the movement of the temperature pins would be ineffective.

Mudge, Cumming, Emery, and other celebrated horologists of the period adapted curbs on the Harrison principle. John Arnold, leaving the spring, decided to compensate for the loss of its elasticity by means of a laminated balance rim composed of brass and steel soldered together, and shaped to a circular form by bending with plyers.

It remained for Thomas Earnshaw to improve the construction. He turned the steel centre and fused the brass around it, thus obtaining a sound junction of the metals and a perfectly circular rim.

The investigations of the two eminent French artists, Ferdinand Berthoud and Pierre Le Roy, were nearly contemporaneous with those of Mudge and Arnold. Berthoud used a gridiron arrangement of brass and steel to compensate for temperature errors, and fitted his timekeeper with two balances geared together. Le Roy experimented with a balance composed of two glass thermometer tubes containing mercury and placed radially, the bulbs, which were filled with alcohol, being furthest from the centre of motion and the tubes pointing inwards.

Berthoud, in 1773, tabulated the effect of temperature upon one of his marine watches. He reckoned that in passing from $32°$ to $92°$ (Fahr.) it lost per diem by—

Expansion of the Balance	62 secs.
The loss of Spring's Elastic Force	312 ,,
Elongation of the Spring	19 ,,
	393 or 6 m. 33 s.

Doubtless Berthoud's observation was correct as far as the total amount of the temperature error goes, but there appears to be no warrant for assuming that a part of the loss was due to elongation of the spring. The thickness and the width of the spring would be increased in precisely the same proportion as the length, and as the strength of a spring varies as the cube of its thickness, the spring would be absolutely stronger for a rise of temperature if the relative dimensions only were considered.*

Sir G. B. Airy, by experiment in 1859, showed that a chronometer with a plain uncompensated brass balance lost on its rate 6·11 secs. in 24 hours for each degree of increase in temperature.

To counteract this effect of change of temperature, chronometers and fine watches are furnished with a balance as constructed by Earnshaw, which expands and contracts with heat and cold. The halves of the rim are free at one end and fixed at the other to the central arm, as shown in Fig. 23. The arm and inner part of the rim are of steel, and the outer part of the rim of brass melted on to the steel. Formerly the brass was preferred to be twice the thickness of the steel. Yvon Villarceau, after investigating the action of the compensation balance, decided that the relative thickness of the metals should be in the inverse proportion of the square roots of their elasticity, giving the thickness of brass to steel as 17 to 12, and lately there has been a disposition to make the proportion of brass to steel as 3 to 2, thus following nearly Villarceau's prescription. Mr. Wright tells me he has found the middle temperature error sensibly less with the thickness of brass so reduced. As brass expands more than steel, the effect of an increase of tem-

* It is curious that Berthoud's statement should have been accepted without question by all authorities and writers till Mr. Wright, the able theoretical teacher at the Horological Institute, pointed out its fallacy in 1882.

perature is that the brass in its struggle to expand bends the rim inwards, thus practically reducing the size of the balance. With a decrease of temperature the action is reversed. The action, which is very small at the fixed ends of the rim, increases towards the free ends, where it is greatest. In a marine chronometer there is one large weight at about the middle of each half rim, which is shifted to or from the fixed end, according as the compensation is found on trial to be less or more than is desired. In pocket chronometers and watches a number of holes are drilled and tapped in the rim, and the compensation is varied by shifting screws with large heads from one hole to another, or by substituting a heavier or a lighter screw. The compensation screws should fit well in the thread and be always screwed home till the heads touch just the rim. If the thread of a screw is at all loose there is danger of distorting the rim by turning the head too tight against it. As a precaution, the shoulders of the screws are sometimes rounded off, so that only the surfaces close to the threads can be in contact with the rim.

In the marine balance there are two screws with heavy nuts on opposite sides of the rim, close to the central arm, for bringing the chronometer to time. These nuts are slit, as shown in the drawing, to clasp the screw spring-tight and so avoid blacklash. In watch balances there are four such

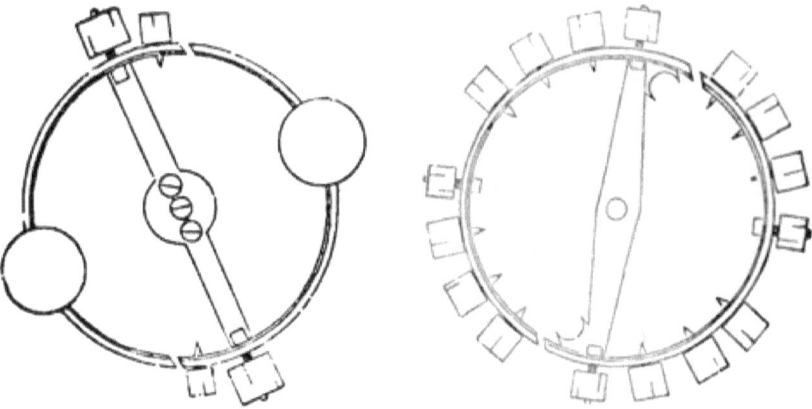

Fig. 23. Fig. 24.
Marine Chronometer Balance.   Compensation Balance for Watches.

screws placed at at equal distances round the rim. These, of course, are not touched for temperature adjustment.

Fig. 23 shows the Marine Chronometer Balance. A compensation watch balance is shown in Fig 24. In all but the finest work the quarter mean-time screws are not fitted with nuts, but made with heavy heads, and screwed into the balance from the outside like the compensation screws. It will be observed that the cuts in the rims of the balances are not radial. The object of cutting them at an angle, as shown, is that the free end of the rim may be stopped from bending unduly towards the centre when the balance is roughly handled.

Before using a compensation balance it is the practice of good adjusters to spin it close to the flame of a lamp, so as to subject it to a higher temperature than it is ever likely to meet with in use. Mr. Arthur Webb raises the temperature till the ends of the rim butt at the notch. The balance is then placed on a cold plate, and afterwards tested for poise. If necessary the balance is trued, and the operation repeated till the balance after heating is found to be in poise, or is rejected. The late A. P. Walsh told me that to get a low temperature after heating he immersed the balance in ether. Mr. Arthur Webb suggests that compensation balances would be more certain and permanent in their action if they were hardened and tempered after the brass is melted on to the steel. It appears probable that uncertainty of action may sometimes be traced to careless hammering of the brass, which is better compressed by rolling than by hammering.

Middle Temperature Error. — The element of irregularity known under this name is one of the most perplexing that troubles the maker of superior timekeepers. The object of applying a compensation balance to a watch or chronometer is that its vibrations shall be performed in the same time notwithstanding that from changes of temperature the energy of the balance spring is varied. In the ordinary form of balance this is sought to be effected by causing the balance to contract with a rise of temperature when the balance spring is weaker, and to expand with a fall of temperature when the balance spring is stronger.

Airy demonstrated that the loss in heat from the weakening of a steel balance spring is uniformly in proportion to the increase of temperature. But the compensation balance fails to meet the temperature error exactly: the rims expand a little too much with decrease of temperature, and with increase of temperature the contraction of the rims is insufficient; consequently a watch or chronometer can be correctly adjusted for temperature at two points only.

While the energy of the balance spring varies equally for equal increments and decrements of temperature, the number of vibrations made by the balance in a given time varies not inversely as the distance of its weights from the centre, but inversely as the SQUARE of the distance of the centre of gyration from the centre of motion. If a chronometer keeping mean time at 60° Fahr. were tried first at 30° Fahr. and then at 90° Fahr., and the compensation weights on the second trial approached the centre the same distance as they receded from it in the first trial, the chronometer might be expected to lose at 30° and gain at 90°. But the difficulty is actually, as already stated, too much expansion in cold, or too little contraction in heat. The action of the balance is complex. The weights do not move radially, and, with an increase of temperature, while the weights are moved inwards the central arm lengthens and the parts of the rim adjacent thereto move outwards.

However as the adjustment can only be perfect at two points, chronometer makers generally arrange so that the compensation is right at the two extremes of temperature which the chronometer is likely to encounter, leaving the greater error at the middle temperature. The chronometer will then be found to gain at all temperatures within the extremes, and to lose if exposed to temperatures outside the extremes.

A marine chronometer is usually adjusted at 45° and 90°, unless special adjustment is ordered to suit particularly hot or cold climates; pocket watches at about 50° and 85°. In a range of 40° Fahr. there would be a middle temperature error of about 2 secs. in 24 hours with a steel balance spring. The amount of the middle temperature error cannot be absolutely predicated, for in low temperatures, when the balance is larger and the oil thicker, the arc of vibration is less than in high temperatures when the balance is smaller and the oil thin; consequently its time of vibration is affected by the isochronism, or otherwise, of the balance spring and the action of centrifugal force on the balance. Advantage is sometimes taken of these circumstances to lessen the middle temperature error by leaving the piece fast in the short arcs.

To avoid middle temperature error in marine chronometers, various forms of compensation balances have been devised, and numberless additions or auxilaries have been attached to the ordinary form of balance for the same purpose. Of these some examples will doubtless be of interest.

The late John Hartnup, director of the Liverpool Observatory, advocated the acceptance of the ordinary compensation balance, and proposed that navigators should be instructed to allow for the middle temperature error, which, after trial of 1,000 chronometers, with steel balance springs, he found amounted to 1·5 sec. in 24 hours for a change of 15° above or below the temperatures at which the chronometers had been compensated. This can only be accepted as an approximation, because, as we have seen, the amount of the Middle Temperature Error is to some extent dependent on other disturbing influences.

With palladium balance springs the middle temperature error is very sensibly less than with steel, which may be accounted for in the following way :—

There is a greater temperature error to be compensated with palladium than with steel springs, and the compensation weights with palladium have consequently to be moved more towards the free ends of the rim. With a rise of temperature the superior expansibility of the brass or outer metal of the rim not only carries the weights towards the centre, BUT, BY ELONGATING, CURLS THE RIM; AND THE PATH OF THE WEIGHTS IS THEREFORE VARIED ACCORDING

TO THEIR POSITION ON THE RIM. It happens that the movement of the weights at the point where the action of the palladium spring causes them to be placed more nearly conforms to what is theoretically required than if they were farther back, as they would be with a steel spring. But the middle temperature error is not yet eliminated or provided for, and its amount, though reduced with a palladium spring, is even less constant than with steel.

Checking the outward movement of the rim in low temperatures, or causing a subsidiary weight to be carried inwards in high temperatures may be taken to represent the two principles on which most auxiliaries are constructed.

At the Guildhall Museum is a watch by John Leroux, who was admitted as an honorary member of the Clock Makers' Company in 1781, and the date mark in the case of this watch (k) seems to indicate that it was made in 1785. If so, it is of especial interest, and may be regarded as the earliest attempt to modify the action of the balance. As will be seen from Fig. 25, the rim is an entire circle of steel,

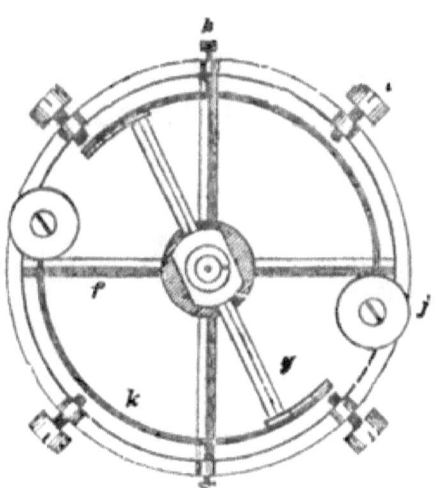

Fig. 25.

with four very light arms. Midway between the arms are left four projections above the upper surface, for receiving the four mean-time screws, as the one marked i; on the upper portion of the balance axis is fitted, friction tight, a moveable steel bar, each end of which carries a circular-lamina, similar to the rim of an ordinary compensation balance, each piece of lamina being firmly attached at one end to the bar by two screws; at the acting end of each lamina is a circular flat weight fixed by two screws.

From the outer rim as before described are two other projections for receiving two small check screws, h; these are made to come in contact with the brass or outer surface of the laminae. Adjustment for compensation is effected by moving the bar forward or backward by pliers applied to its flatted sides. This motion of the bar causes a different part of the laminae to bear against the screws, h, thereby lengthening or shortening the effective compensation length.

Molyneux's auxiliary (Fig. 26) is attached by a spring to each end of the central arm, and is acted on by the free ends of the rim in high temperatures only. A screw in the end of the rim and another in the auxiliary serve to adjust the action as may be required. Molyneux's patent also covered the use of a short laminated arm instead of the spring by which the auxiliary is attached to the central arm, and many successful auxiliaries are now made in that way, of which Fig. 27 may be taken as an example. The laminated auxiliary arm is generally about one half as wide and one

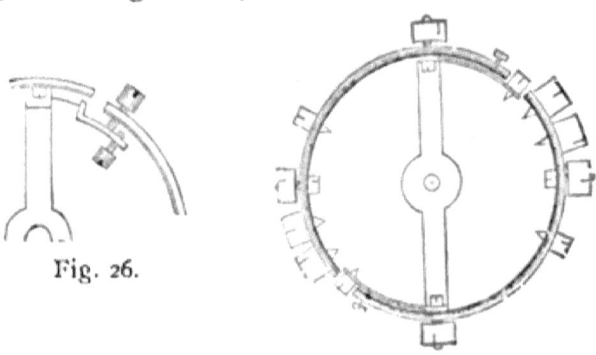

Fig. 26.

Fig. 27.

quarter as thick as the rim of the balance, and it can be made to act either in high or in low temperatures according as the brass is placed inside or outside of the arm. Some chronometer makers consider there is an advantage in constructing the auxiliary for low temperatures, because compensation balances, after exposure to heat, are found to set slightly inward instead of returning exactly to the normal position; and, while an auxiliary acting in high temperatures would add to the error when such setting occurred, one acting in low temperatures would tend to neutralize it.

John Poole's best-known auxiliary (Fig. 28) consists of a piece of brass attached to the fixed ends of the rim, and carrying a regulating screw, the point of which checks the outward movement of the rim in low temperatures. Poole also used a butting screw, fixed so as to stop the outward movement at the free end of the rim; and a more recent inventor met with success by connecting the two free ends of the balance rim with a thin wire, and so checking their outward movement in low temperatures.

William Hardy attempted to avoid the middle temperature error without using an auxiliary by altering the form of the balance. Abandoning the cylindrical laminæ used by Arnold and Earnshaw, he used a straight laminated bar of brass and steel, the brass being underneath (Fig. 29). A hole in the centre of the bar served to attach it to the

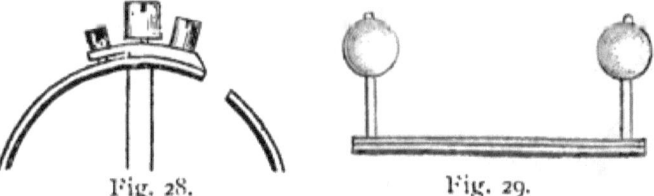

Fig. 28.   Fig. 29.

staff, and at each end of the bar was a stalk carrying a spherical weight. These weights could be made more or less active as compensators by screwing them up or down on the stalks which had threads cut on them. By slightly curving the laminated bar upwards or downward, the weights could be made to approach or recede from the centre of the balance more or less as desired. At first sight it appeared that the difficulty of the middle temperature error had been overcome. But to obtain sufficient compensation the laminated bar must be so thin and the stalks so long as to leave the balance wanting in rigidity. Nevertheless,

Hardy's attempt led to the invention of many other balances on the same principle.

Massey, in 1814, patented (No. 3854) a balance very much resembling Hardy's in appearance, but in which the central arm was solid, and the upright stalks formed of brass and steel. On a central arm, similar to Hardy's, E. J. Dent mounted laminated pieces bent over like staples thus ⊏ to increase the action.

In 1849 John Hartnup invented the balance shown in Fig. 30. The rims are composed of brass and steel, as usual, but they are neither upright nor flat, but bevelled, or placed at an angle midway between these two positions. The central arm a is also bimetallic, the brass being uppermost, and connecting the arm with the sections of the rim are two other bimetallic strips $b\ c$, the brass of these being underneath, and the steel on top; $e\ e$ are the weights, and at the ends of the rim the screws for timing and poising.

Subsequently Victor Kullberg constructed a flat-rimmed

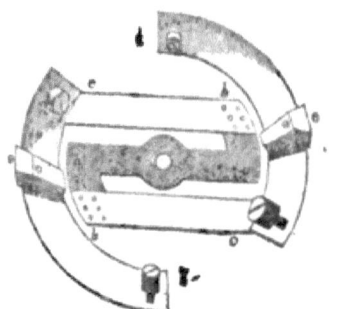
Fig. 30.—Hartnup's balance.

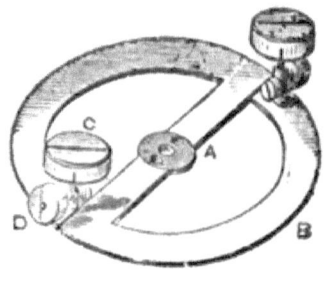
Fig. 31.—Kullberg's balance.

balance, as shown in Fig. 31. Here the central arm A, and the rim B, composed of brass and steel, are in one piece, but in the arm the brass is on top and in the rim underneath, so that with a rise of temperature the ends of the arm bend down and the free ends of the rim are lifted upwards and inwards. The weights C are carried on stalks, which also afford a support for the timing nuts D.

E. T. Loseby's Balance.—Chronometers with this balance were remarkably successful at the Greenwich trials from 1846 to 1853. The rims are bimetallic of brass and steel, shorter than usual; at the end of each segment of the rim is a cup-joint in which is placed a glass vessel consisting of a curved arm and a bulb, which contains mercury (Fig. 32). The curved arm is sealed with a little air in it to ensure the continuity of the thread of mercury when it contracts. It is apparent that by bending the cup-joint the direction of the glass arms may be altered, and in this way a very exact temperature adjustment obtained.

Loseby's invention was admitted by the Greenwich authorities to be an improvement, but his application for a reward was refused, the Admiralty, as a sort of recompense, buying a larger number of his chronometers than they otherwise would have done. This rather shabby treatment disgusted Loseby, who gave up chronometer-making and

Fig. 32.—Loseby's balance.

returned to Warwickshire, where he died in 1890; but the manufacture of balances on his principle has been occasionally revived, with, I believe, encouraging results.

The late C. van Woerd, of Waltham, Mass., patented an ingenious method of construction. Across the periphery of each half of a steel rim, for about a third of its length, near the fixed end, he formed V-shaped notches. Brass was fused on to the steel and turned off level with the circular part of the steel rim, leaving it bare except at the notches, which were thus filled with prisms of brass. By varying the number of notches he claimed to be able to obtain any desired amount of movement of the free ends of the rim.

CHAPTER IV.

Method of Procedure in Springing and Adjusting.

Applying a flat spring.—The number of vibrations required depends, of course, on the train. For a modern watch with a seconds train in which the fourth wheel turns once in a minute, divide the number of the fourth wheel teeth by the number of leaves in the escape pinion; multiply the quotient by 30 (double the number of escape wheel teeth), the product will be the number of impulses the balance receives in a minute. If it is an 18,000 train the number will be 300, that is 5 beats a second. A 16,200 train gives 270 a minute; 4·5 a second.

Should the watch not have a seconds train we must go back to the centre wheel which rotates once in an hour. In this case multiply together the numbers of centre wheel, third wheel, fourth wheel teeth and 30. Also multiply together the number of leaves of third pinion, fourth pinion, escape pinion and 60. Divide the first sum by the last and the quotient will be the vibrations per minute.

Having ascertained the number of vibrations required proceed to select what appears to be a suitable spring. Lay the spring with its centre coinciding with the cock jewel and mark on the coil that is to be the outer one exactly where it would enter the stud hole, bearing in mind that the spring should be rather small in diameter than large; FOR A SPRING TOO LARGE IN RELATION TO THE INDEX PINS AND STUD IS PRETTY WELL SURE TO BE SLOW IN THE SHORT ARCS.

The next thing is to count the vibrations of the balance when connected with this spring.

Put the eye of the spring over the balance staff down on to the balance, and press the collet on to the spring so as to confine the inner portion that will be broken away for the eye. If it should happen, as in rare cases it may, that the collet when pushed on the staff its proper way fails to hold the spring, the collet may be reversed and pushed on with the smaller side of the hole first.*

Place in a convenient position on the bench a watch that is known to be keeping correct time, then take hold of the spring firmly with tweezers at about the distance of the curb pins short of the spot marked for the coil to enter the stud; then lift the spring so that the balance hangs horizontally and the lower balance pivot is just above the watch glass. Give the balance about half a turn so that it will vibrate for over a minute, Owing to the contraction and expansion of the spring the balance will also acquire an up and down motion.

With your eyes on the watch dial count every alternate vibration registered by the contact of the pivot with the glass till a spring is obtained that gives 75 double vibrations, if it is for an 18,000, and from 67 to 68 during half-a-minute, if for a 16,200 train. The observation may extend over a lesser or a greater period, always remembering that for an 18,000 train, there are to be 5 DOUBLE vibrations every TWO SECONDS, and for a 16,200 train $4\frac{1}{2}$ in the same time. It is better for the spring to give say half a vibration less in a minute rather than more, for it is sure to be a little faster when the eye of the spring is broken out to suit the collet. Extend the observation and counting to one minute if any doubt exists respecting the watch which is taken as a standard. It should not only be a good timekeeper but also have the seconds dial correctly spaced, and the seconds arbor concentric therewith if exact results are to be obtained for shorter periods than a minute. Of course, if a chronometer is available it may be used instead of a watch. Many prefer to listen to the taps of the pivot on a watch glass and to watch the seconds hand of a regulator. Others will

---

* In the absence of the collet a bit of beeswax or putty powder the size of a small pin's head may be placed upon the end of the lower balance staff pivot to catch the eye of the spring under, in order to count the vibrations; but the use of wax or powder, a portion of which may be afterwards transferred to the watch, is not to be recommended. Besides, clasping the spring between the balance and the collet places that part of the eye which will be broken off out of action, and the result will be more exact than catching up the spring close to its centre.

D

dispense altogether with the audible report on a watch glass and catching up the spring will give a slight twist of the hand holding the tweezers, and watch the number of vibrations given during 4 or 5 seconds by listening to the beats of a regulator.

Yet another plan is to hold to the ear a watch having the same number of vibrations as it is desired the balance under trial should have, and note the discrepancy, if any. The first method given is, I think, best suited to a tyro, and will, I am sure, result satisfactorily if carefully conducted.

Although very accurate results may be obtained by "vibrating" on the watch when continued for, say, two minutes, the use of a "Vibrator" is recommended by Mr. Geo. Walker. With it there is certainly the advantage that the attention may be withdrawn from the balance under trial for a few seconds without the observation being lost.

The "Vibrator" consists usually of a vertically pivoted box, with glass top, containing a balance and spring accurately adjusted to give a standard number of vibrations. If adjusted to, say, 18,000 vibrations, another small box is provided to give 16,200, and sometimes a third giving 14,400. Any of the two latter may be slipped over the first when these numbers are required. Attached to the side of the pivoted box there is a vertical pillar from which projects a horizontal arm carrying tweezers to hold the outer end of the spring. The construction of the tweezers is such that the spring may be drawn out or in by simply turning a small milled nut.

The balance and spring to be tested are adjusted over the standard balance, the outer end of the spring being held by the tweezers so that the bottom pivot remains in contact with the glass throughout the entire vibration. The two balances are then set in motion, exactly together, by a push from a lever pivoted to the stand. With a little practice it is easy to quickly adjust the length of spring so that the balances are running accurately together.

For the drawing of the Vibrator (Fig. 32A) I am indebted to M. Faure.

When a spring of the right strength is found it may be

pinned to the **collet**. To **break** away sufficient of the eye for it to pass over the collet place the spring on the board

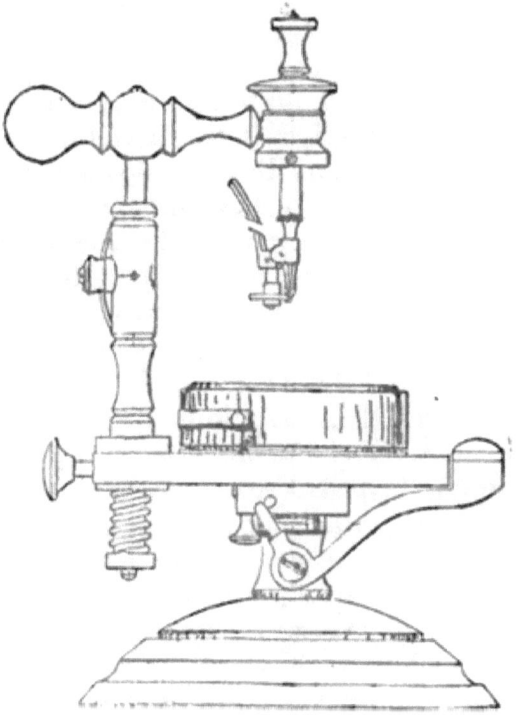

Fig. 32A. Vibrator.

paper; grasp it with tweezers at the point where it is to be severed; with a screwdriver, or other suitable implement, bend the eye close to the tweezers, and it will break away.

Assuming the position of the stud to be fixed, and that the spring should start from the collet on a level with the centre of the cock jewel if the watch were on edge with the pendant uppermost, place the spring on the cock, with the mark already made for the outer attachment coincident with the face of the stud, and mark on the eye of the spring the point where it is to spring from the collet. The spring should start away from the collet with an easy curve, and must not hug the collet, or isochronism will be out of the question. With tweezers, having a rounded grip rather than a sharp edge, hold the eye a little farther up the spring than the mark just made, and with a peg gently bend round to the requisite angle the piece which is to enter the collet. Do not press with the peg close to the tweezers, or the spring may break. A sketch of the **collet showing** the hole for the spring may be made on the **board** paper as a guide for the angle. The piece bent round, of course, is curved. Hold it in the tweezers, and, with the assistance of the peg, straighten it.

Occasionally it may be that the only spring of the required strength available is too small with relation to the stud. If unhardened a spring may be expanded by placing it on a flat plate previously warmed, and heating it to a bluing temperature. It must not be heated beyond a bluing temperature, but the operation may be repeated more than once, and the spring will expand at each heating. To ensure the flatness of the spring remaining undisturbed it may be covered by a small flat piece of brass or **glass**, which must not be too heavy, **or it** may prevent **distention** of the **spring.**

The Collet.—The depth of the collet should not be less than twice the diameter of the central hole, if room allows. The taper of the hole and the taper of the balance staff where it fits should exactly correspond, so that the collet does not rock, and that part of the staff should be left grey to allow a tighter grip for the collet. The bottom of the hole should be rounded, so that it readily enters on the staff; the bottom outer edge of the collet should also be rounded, to facilitate the entrance of the tool used for raising it. The hole for the reception of the spring must **be** exactly at right angles to the central hole, for if the spring has to be forced or bent to suit the hole it is hopeless to expect good results. Beginners are often in such **a** hurry to pin the spring on that they proceed when **the** hole is obviously not in the proper direction. The end of the spring should be passed into the hole, **just kept in** position with **a** temporary **pin and examined.** If any alteration is **needed, the spring is to be taken out and** the hole broached **in the direction required till it is right.** A round pin of **hard brass is then fitted to the hole.** When this is done, take a waste piece of the spring, insert it in the hole, and flatten away the pin till it and the spring together fill the hole properly. Then cut off the pin, leaving but little to project, so that there is no danger of its touching the spring in action.

The usual plan of pinning in is to put the collet on a broach held between the thumb and finger of the left hand while the pin is fitted and the spring pinned on, taking the precaution to push a piece of paper on the broach before the collet or the spring may touch the fingers, and in the case of a damp hand the spring would be likely to be spoilt by rusting. By previously fitting the pin as directed, there is now very little fear of injuring the spring. The pin can be pressed home with a small joint pusher.

Another mode of pinning on is to place the collet on the board paper, adjust the spring to the collet, and with a short piece of boxwood sloped away at the end press the collet on the board; the pin can then be fitted with comfort, and without danger of shifting the collet.

Yet another way of proceeding to pin the spring to the collet is to get a plate of brass about an inch square, and tap a hole in the middle of it less in size than the hole in the collet. A screw passing through the hole in the collet fixes it to the plate. The advantage of this plan is that the serface of the plate serves to show at once if the spring lies in the proper plane. Care must be taken with the underside of the head of the screw, so that it does not scratch the collet.

The collet may now be put on an arbor, and the arbor rotated in the turns to ascertain if the spring is true. In setting the spring it must only be touched close to the eye. Steady timekeeping will be out of the question if the spring is bent to and fro in reckless attempts to get it true. The eye should be brought gradually round to get it in circle, taking care not to overdo it. When this is right, and the spring is also true on the face, many good timers heat the spring and collet to a blue to set the eye. The outer coil may now be pinned into the stud.

It will be prudent to first pin the spring in temporarily

and notice if the eye is true with the cock jewel, or very slightly towards the stud. If it is not, and the stud cannot be shifted, the stud hole had better be broached in the required direction. Or, of course, the spring can be slightly bent close to the stud hole to bring the eye right with the cock jewel. The largest part of the pin should be towards the body of the spring, and therefore the hole should be broached from that direction. First fit a round pin, and then flatten it to suit with a waste piece of the spring as directed when pinning to the collet. The pin at the stud may be left fairly long. The waste outer part of the spring should not be broken off close to the stud, but sufficient should be allowed to project through the stud to allow for any letting out of the spring that may be required.

The curb pins should be of brass, only just free of the spring, parallel inside and tapered back from the point on the outside, without burr or roughness that would allow a lodgment of the spring. Mr. Kullberg considered the curb pins should not reach below the bottom of the spring, so that there is no room left for the second coil to jump in. The curb pins should not extend far beyond the stud, or the outer coil may jump out. But apart from the jumping of the spring it is objectionable to have the curb pin or pins far from the stud. It is noticed that watches in which this distance is great are very difficult to time.

The watch should be tried, lying and hanging, for twenty-four hours in each position. If it gains in the short vibrations the spring may be taken up, and if it loses in the short vibrations let out a little. But if the alteration has the opposite effect to that desired, as in some instances it may, proceed in the contrary direction.

If the short arcs are slow, some recommend bending the outer curb pin away from the spring and pressing the inner one hard against it, so that the outer coil is really dragged out of its position. This seems to be a most objectionable nostrum, which doubtless originated through a spring too large having been applied.

If the short arcs are fast the curb pins are sometimes slightly opened as a remedy, but this is also open to objection, for the pins then obviously cease to be parallel with each other, and with undue play between the curb pins the spring will get worn where they touch it, and the going of the watch will not be satisfactory.

Unless there is some grave constructional fault the variation will not be so excessive but that the watch may be brought to time with the index, dividing the error between the long and short vibrations.*

Where there is no seconds hand, the usual plan is to make a mark on the fourth wheel; but Mr. Bickley recommends placing the watch to the ear and counting every alternate beat for a minute or two, and comparing with a regulator or watch going to time. However, a small dot on the fourth wheel is an almost imperceptible disfigurement, and will generally be found useful as the watch is getting pretty close to time.

Macartney's Collet Adjuster. — For twisting a balance spring collet to position, the special tweezers

Fig. 33.

shown in the sketch (Fig. 33) are admirable. One jaw is concave, and the other in the form of a knife edge.

Plose's Collet Lifter. — This is an excellent form of pliers for raising a balance spring collet. The jaws are curved and have knife edges. The leverage is much more even and effectual than with the old style by one-sided wedging. There is a set screw to prevent the edges closing too far and so injuring the staff (Fig. 34).

---

* All watch balances should, I think, be furnished with screws for ready adjustment of weight. The extra cost, even for cheap watches, is not worth consideration. There is no doubt that the proper way of bringing the piece to time is by varying the balance instead of lengthening or shortening the spring the the stud, which upsets any prearranged position for the inner point of attachment.

For pressing a collet home, a table with holes or slits such as is depicted under the head of "Stud" will be found useful.

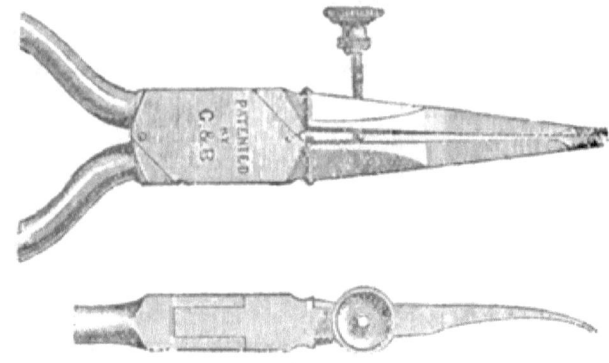

Fig. 34.

Balance Spring Holder.—This tool (Fig. 35) forms a kind of self-acting tweezers, and is useful for holding the outer coil of the spring, especially when it is desired to get

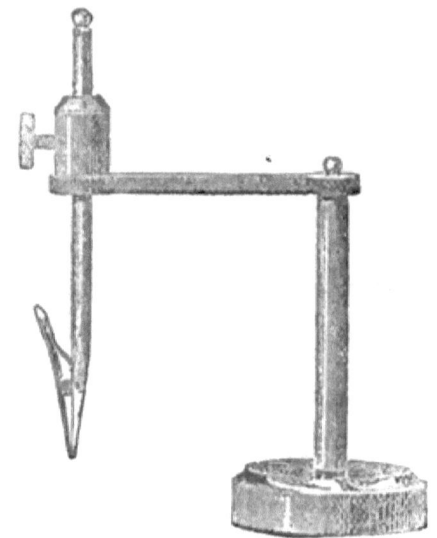

Fig. 35.

the length accurately with the balance in its place in the watch before pinning to the stud, as in the case of Bréguet spring prior to turning the overcoil.

Breguet Spring, or Flat Spring with overcoil.—
The selection of a spring, counting its vibrations with the balance and pinning to the collet may be conducted in the same way as for a spring without overcoil, except that if it is decided to keep the point of attachment to the collet in a horizontal line with the cock jewel when the watch is placed pendant up, allowance must be made for the bending in of the spring to reach the stud, which, of course, is nearer to the centre than a stud for a flat spring would be. Experienced springers after selecting a suitable spring, by vibrating, break away the eye very accurately, to ensure the desired result. The overcoil will not materially affect the number of vibrations in a given time. It is, however, difficult to frame a rule of any practical use to the beginner, because there are so many factors to be considered; the diameter of the spring, the length of the overcoil, the distance of the stud from the centre, the thickness of the wire from which the spring is made, all affect the amount to be broken away at the eye. To be quite certain the spring selected will bring the watch to time after the overcoil is formed, it may be desirable to try it in the watch, and for this purpose the holder (fig. 35) will be useful for gripping the outer coil.

The overcoils of watch springs are usually turned with curved nosed steel tweezers like Fig. 37. These curves

Fig. 37.

should be of suitable shape for the particular spring, and be well polished. Beginners are not sufficiently careful in forming the curved noses. When a bit of spring wire is gripped between them they must show an exactly circular outline. If defective they would twist the spring, even in the hands of an experienced operator. In forming the Phillips' curves, some watchmakers use hot pliers of the requisite shape to set the curve to the required form, but this is exceptional treatment, and does not appear to be necessary.

Fig. 36 is a useful little clamp of brass, by Mr. T. D. Wright, for holding a spring while turning the overcoil, the jaws of which are curved to the outer coil of a spring of average size. The upper screw passes through a clearing hole in the loose jaw and is tapped into the body of the tool; the pin below is a steady pin; the bottom screw tapped into the loose jaw serves to keep the opening between the jaws parallel to suit the thickness of the spring; a is a screw, and b a steady pin for attaching it to an old watch plate or other table.

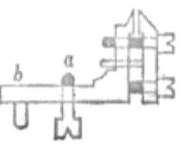

Fig. 36.

With this tool, **and care,** the turning of the overcoil will be comparatively easy. Having decided on the length and **form** the overcoil is to take, and of which you will have **either a** pattern or a **photograph, or** a drawing on paper or **metal,** break off a piece **of the waste outer** coil and bend it **to the shape** required; **then pull it out again to** coincide with **the outer coil, and from the point where the spring** enters **the stud,** or **if there is to be an index, the middle** position **of** the curb pins, mark off **the length of the curve.** Then fix **the** outer coil in the **clamp just short of the** point where the overcoil should start, **and place underneath** it **a** sketch of the overcoil for guidance. There are now two distinct operations—first, to raise **the** overcoil above the plane of the spring, and then to form the curve. If the two are attempted at once, it is most likely that in unskilled hands the spring will be twisted and practically spoiled. Therefore with strong tweezers raise the spring sufficiently, and then with the curved-nosed tweezers bend the curve very gradually without twisting the wire, taking care not to overdo it in any **part, for bending to and** fro **is sure to** deteriorate **the spring.**

Presuming that the copy be a theoretically isochronal curve, let the arch of the overcoil, if it deviate at all, be rather nearer the centre than the other way, so as to be, if anything, quicker in the short arcs. When the curve is **correct it** may be set by heating the spring to a bluing temperature. Grandjean **suggests gently heating the spring** on a white enamel dial till a drop of oil placed alongside the spring begins to smoke.

If the spring is to be free, without an index, pinning to the stud is a simple affair, requiring no instruction beyond what has already been given when dealing with a flat spring; but with an index there are two methods of procedure, one of which is to be recommended. The plan formerly most generally followed was to carefully shape that part of the outer end of the curve where the index pins would operate into a circular form of a radius equal to the distance of the curve from the centre of the index; that is, from the centre of the cock, and then proceed to pin to the stud. But in almost every case this last operation distorted the circular part on which pains had been bestowed, and the shaping of this had accordingly to be done over again. The best plan is to pin to the stud first, and then shape the circular part to suit the path of the index pins. This way is more expeditious, and the spring is subjected to less bending.

If the balance spring has been carefully selected the watch will not show a considerable variation from mean time. Leaving the final adjustment to be made by screwing the mean time screws in or out, it may now be brought to within about two minutes a day by, if necessary, adding to or taking weight from the balance. Usually two opposite screws are removed and heavier or lighter ones substituted, as may be required. Platinum screws are used when extra heavy ones are needed. The overcoil of the spring should not be touched for mean time adjustment.

Possible tests and corrections now to be made may be classed under the heads of:—

        Isochronism.
        Compensation for temperature.
        Position errors.
        Mean time adjustment.

There is no absolute rule as to the sequence of these

operations. Some adjusters proceed in one way while others, equally successful, pursue a somewhat different course. Isochronism is, as a rule, dealt with first, because it is possible more easily to carry it to completion in presence of errors coming under the other heads enumerated. Much, however, depends on the character of the watch, and the performance required from it. Between the straightforward procedure adopted with a medium class piece and a fine watch which has to be subjected to the exacting ordeal of an Observatory test there is a wide difference. In the latter case the margin of error allowed is so narrow that perplexing constructional inaccuracies, quite distinct from the balance and spring, have to be located and rectified.

We will first take, as an example, a going barrel watch not required to be adjusted closely for positions.

Rack or Holder for Watches on Edge.—The primitive and most usual rest for watches placed on edge for positional adjustment consists of a rack or rest of wood having V-shaped notches to suit various sized watches.

A convenient movement holder sometimes used is shown in the annexed cut (Fig. 38). It is adjustable, and capable of being turned in any desired way for position timing.

Timing Box.—A brass box should be provided for the reception of an uncased watch movement while it is being timed. The edge of the pillar plate rests in a rebate, and a cover with a glass let into the top screws on to the box, and keeps the movement in position.

Fig. 38.

Isochronism Adjustment.

Tests for isochronism may extend over different periods, but variations are always reduced to 24 hours for comparison. Although approximations may be first obtained by observations at short intervals, 24 hour trials should, if possible, be allowed as the rates become close. Some adjusters check the watches just when they happen to be at leisure, and record the duration of each trial in minutes, reducing the variation to 24 hour periods by the aid of logarithms, or a slide rule to save extended calculation. Aliquot parts of 24 hours can, however, generally be managed by systematic procedure, and there is then less liability to error from miscalculation. Trials of less than four hours' duration can scarcely be regarded as a reliable guide.

Dial up is generally taken to represent the long vibrations, and the mean of two opposite vertical positions the short vibrations. With short trials of going barrel watches the same turns of the mainspring should be used for each test of the same kind. If we begin with four hour trials it will be convenient to start in the morning, to wind the mainspring but one turn so as to get the shortest vibration the balance will have in ordinary use, and to set the watch to mean time or in agreement with a reliable regulator. After four hours pendant up the watch is found to have lost three seconds, and its variation from mean time —3 is noted. The mainspring should then be wound about half-a-turn, and the watch placed pendant down. After four hours running it is found to be —8, having lost 5 secs., and its variation from mean time —8 is noted. The mainspring may then be fully wound to get the longest vibration, the watch placed dial up and allowed to run for four hours; it is then found to be —4, having gained 4 secs. The mean variation of the two positions representing the short vibrations is represented by — 4, and the variation in the long vibrations by + 4; as 4 hours, the duration of each trial, are one-sixth of 24 hours each of these results is multiplied by 6 to obtain the variation in 24 hours, giving respectively — 24 and + 24, which added together = 48; the short arcs are said to be 48 secs. slow.

If the variations had been plus in each case or minus in each case one total would have been subtracted from the other, and the remainder would have represented the daily rate.

The most usual way of making the long and short vibrations isochronous is by altering the form of the overcoil. When the short arcs are slow, as in this case, the arch of the overcoil is closed slightly, and a little more of the body of the spring added to the overcoil. When the short arcs are fast, the arch of the overcoil is made a trifle flatter, and a little of the overcoil taken back into the body of the spring. Fig. 39 shows clearly what is meant. If the full line represents the original form of curve, it would be altered in the direction of b to quicken the short, and in the direction of a to quicken the long arcs. To alter the shape of the overcoil the spring is held with a pair of tweezers with brass faces curved exactly to suit it, while the alteration is made with a similar pair. Alterations must be made very gradually, for if bent too much the spring is likely to be spoilt in bending it back.

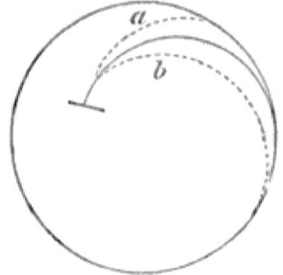

Fig. 39.

After the requisite alteration is made, the watch may be again set to mean time in the morning with the mainspring fully wound, and after 24 hours running, dial up, it is found to have gained 35·5 secs., this variation + 35·5 secs. having been noted, the watch with about two turns of the mainspring wound may be tried 12 hours pendant up; it is then found to be + 50·5 secs. fast, having gained 15 secs; this is noted; the mainspring wound about a turn and a half and the watch placed pendant down for 12 hours; it is then found to be 59·5 secs. fast, having gained 9 secs. The short arcs are now slow, at the rate of 11·5 secs. in 24 hours; the balance spring therefore requires a slight alteration in the same direction as before.

After this has been done the watch is fully wound, again set to mean time and after 24 hours' trial, dial up, is found to be, say, 36 secs. fast. The watch is noted + 36 s. With the pendant up for 12 hours, it gains 21 s., making it + 57 s. And with the pendant down 12 hours, it gains 15 s., and is then + 72 s. So that the variation in the long vibrations is + 36, and the sum of the variations in the short vibrations is + 36 also. The watch is now said to be isochronous, but there is still a position error between pendant up and pendant down of 6 seconds in 12 hours.

Twelve hours' trial with pendant to the right shows a gain of 27 s., and 12 hours' trial with pendant to the left a gain of 13 s., making a position error of 14 seconds in 12 hours in the quarter positions.

The mean time screws may be drawn out sufficient to give a loss of 38 seconds in 24 hours, thus dividing the error. The piece would then show a variation from mean time in 24 hours running in each position :—Dial up—2 s.; pendant up + 4 s.; pendant down — 8 s.; pendant right + 16 s.; Pendant left — 12 s.

This is not at all a bad result, and the watch being a little faster pendant up than dial up is likely to approach still nearer in these positions as the oil thickens. We may, therefore, leave it as sufficiently near.

Position errors, which are due to escapement faults, bad jewelling, and other constructional inaccuracies, are often confounded with a want of isochronism; but a watch may be perfectly isochronous, and yet have very large position errors.

If the short vibrations are not more than a turn in extent, the vertical positions could be equalized by drawing out the quarter screws of the balance nearest the fast position, and setting in the ones nearest the slow position, and this is the remedy generally adopted. If the short vibrations are a turn and a quarter, altering the screws will be useless. Should the short vibrations be a turn and a half, which is not very likely, the opposite method would have to be resorted to, viz., the screws opposite the fast positions must be set in, and those opposite the slow position drawn out. This is called "timing in reverse." The best authorities strenuously object to tampering with the balance, and prefer to leave the position errors alone.

No doubt it is a grave fault to set the balance out of poise, and as the extent of the vibration falls off from thickening of oil and dirt the correction is destroyed and an additional error introduced. Of course, with a first-rate watch, where everything is of the best, the position errors will be very small.

There is another method employed by some to lessen the position errors, but it is not less objectionable, and is certainly not sufficiently efficacious to be worth consideration if the position errors are large. It consists of bending the balance spring (not the eye), so that the eye, instead of being true with the cock jewel, presses against the staff in the fast positions; in the preceding example, the spring would be moved from the pendant, and to the right of it. The effect of this is to cause more friction on the balance staff pivots at those points.

A watch out in positions may often be brought nearer by turning the roller of the escapement half-way round on the balance staff, slight defects in the pivots and holes being thus neutralized.

There should not be much difference whether the short vibrations are caused by decrease of motive power or by change of position from lying to the rack, yet the results will not be absolutely identical. If the discrepancy is very great, it would probably indicate defective jewelling, holes too long or too large, or not sufficiently polished. The length of the straight part of the jewel hole should be about equal to its diameter.

In adjusting marine chronometers and chronometer clocks for isochronism, the long and short arcs are not obtained by changing the position of the instrument, but the arc of vibration is reduced about one-quarter by letting down the mainspring, or reducing the power in any other convenient manner. For instance, a stiffish spring may be brought to bear on the fourth wheel pinion with sufficient pressure to absorb as much of the force as may be desired.

The method of recording observations varies with different adjusters. The subjoined arrangement appears to be convenient. Symbols are used to indicate the the positions follows:—Dial up = O; dial down = O; Pendant up = T pendant right = ⊣; pendant left = ⊢; pendant down = ⊥. The time is set down in hours and quarter hours.

Date.	Mainspring, how wound.	Time of day trial begins.	Watch time.	Position.	Time of day trial ends.	Duration of trial.	Watch time.	Daily rate.	Result. What alteration.
Oct. 2	1 turn	8a	R	T	12	4	−3 ⎫	−24	Shorts 48 secs. slow. Spring set.
,,	,,	12	−3	⊥	4p	4	−8 ⎬		
,,	Full	4	−8	O	8p	4	−4 ⎭	+24	
3	Full	7¾a	R	O	7¾a	24	+35·5	+35·5	
4	2 turns	7¾a	+35·5	T	7¾p	12	+50·5 ⎫	+24	Shorts 11·5 secs. slow. Spring set.
,,	,,	7¾p	+50·5	⊥	7¾a	12	+59·5 ⎭		
5	Full	8a	R	O	8a	24	+36	+36	Set to mean time.
6	3 turns	8a	+36	T	8p	12	+57 ⎫	+36	
,,	,,	8p	+57	⊥	8a	12	+72 ⎭		

Adjustment of a Fine Watch.—Where the best possible performance is to be obtained from a watch in every position and in varying temperatures an exact standard of time for comparison is essential. Without this it would be foolish to undertake the task. Even with an Observatory time signal at frequent intervals there should be two good regulators to check a doubtful signal before making any alteration to the watch as it is approaching completion.

However, assuming all the necessary apparatus to be at hand the whole of the mechanism of the watch should undergo a rigorous scrutiny prior to proceeding with the timing and adjustment. It is often the next to nothing, the very slight errors of construction which, if overlooked, give trouble at the end. Taking, as before, a going barrel watch with a lever escapement we will suppose that an overcoil balance spring has been applied as already directed. The escapement should then be examined in accordance with the procedure given in Chapter IX., p. 132.

The escapement having been passed as correct, and the

mainspring fully wound the balance should have a vibration of not less than a turn and a half nor more than a turn and three-quarters with the dial up, and not less than a turn and one eighth with the pendant up.

If with the pendant up the arcs are only a turn in extent a stronger mainspring should be tried, and this remedy will probably prove effective; of course, if a balance too large or heavy has been selected a fresh balance must be obtained and the watch resprung.

Assuming the vibration to be satisfactory, the watch may be brought somewhere near to mean time—say, to within 2 minutes a day—by, if necessary, adding to or taking weight from the balance.

Bonniksen recommends springing the watch so that there is at this stage a loss of four minutes a day to allow for weight to be taken from the balance in poising it, but this is not the usual practice.

Then the compensation should be proceeded with and almost perfected. If tests for isochronism were attempted first variations due to changes of temperature might easily be mistaken for want of isochronism. The compensation cannot be quite concluded at this stage, because, subsequently poising the balance may slightly derange it. On the other hand it is useless to think of finally poising the balance now, because re-arrangement of the screws, and a possible set of the balance during the compensation adjustment will probably affect the poise. But the fact of the balance being untested for poise renders it necessary to provide *perfectly level* surfaces in the refrigerator and oven on which to lay the watch so that want of poise will not affect its rate.

With the compensation correct, or nearly so, try the balance for truth in the callipers or on the poising tool as directed in the chapter devoted to the subject; get it as perfect as possible and then proceed finally to test the poise of the balance and spring in the watch. The mainspring should be wound but little so as to give only a full turn or rather less when pendant up. It is very rarely that the vibration cannot be sufficiently reduced if the stop work is removed. In an extreme case it may be necessary to substitute a weaker mainspring.

Suppose the watch when horizontal gains at the rate of 30 seconds a day. Try it pendant up; its rate is then $+36$. Pendant down it is $+16$.

It may be assumed that the balance is too heavy at the bottom when the watch is pendant up and as a ready improvement for a watch which is not to be adjusted in positions the timing screws nearest this position are proportionately screwed in, and the opposite ones similarly withdrawn.

But clearly the heaviest part of the balance may not be exactly opposite the pendant, therefore let us try the quarters:—

We find with pendant right the daily rate is $+32$.

,, ,, pendant left ,, ,, $+21$.

Probably the quarters would not be so close on account of constructional errors and the possible presence of these renders it unsafe to accept the deviation as due wholly to want of poise in the balance.

Therefore wind the watch fully so that the vibration is quite a turn and a half, and try round the vertical positions again.

Pendant up, its daily rate is $+40$.
,, down, ,, ,, $+20$.
,, right, ,, ,, $+35$.
,, left, ,, ,, $+25$.

It is again hardly likely there would be no other disturbing elements, but accepting these figures as an example we may ascertain if it is an open face watch that the balance when at rest is too heavy between the IV. and the V. This may be adjusted by screwing in a timing screw or nut at this point or by proportionally screwing in the adjacent ones.

ABSENCE OF ISOCHRONISM, AND VARIATIONS BETWEEN HANGING AND LYING. SUMMARY OF CAUSES AND REMEDIES.

The natural tendency of the escapement which causes a loss in the short arcs.

Various escapement faults; excess of shake in pivot

holes, or insufficient freedom in pivot holes; escape wheel too heavy; bad train depths; escapement out of beat.

Variations between dial up and dial down.—Balance spring badly pinned to collet. Any twist in the spring here is almost sure to cause variation in these positions. Probably the hole in the collet is not exactly in the proper direction, and the best plan is to have a new collet.

Want of uniformity in escapement end shakes, causing the end of the impulse pin to graze the lever of the safety finger; or the balance to graze the escape cock.

If the impulse pin is not perfectly upright it may in one of these positions be free as it passes in and out of the horns, and in the other positions may graze.

A short balance-staff pivot may in one position just rest on its conical shoulder.

A defective or unevenly set end stone.

Excessive or careless oiling may cause a drop of oil on some part of the escapement to spread and create sufficient adhesion in passing an adjacent surface to retard the balance and diminish its vibration. See especially that contact between the lever and banking pins is free and not sticky from oil.

Variations in the Quarters, Positional Errors.—Want of poise in the balance.

Want of poise of the lever and pallets.

Too little draw, causing the safety pin or finger to rub on the roller; this may occur only in one position, owing to want of poise in the lever and pallets.

Extra Long Arcs defective.—If the extra long arcs are fast, the fault may be that the impulse pin knocks the outside of the lever. To correct this a weaker mainspring may be substituted for the existing one.

If the balance spring is pressed out of its proper position by one curb pin, though there is play between the two, it may be that the spring leaves the curb pin and becomes free in the extra long arcs, causing a loss.

The balance when expanded to its fullest extent through natural action in low temperature and by centrifugal force may just touch some part of the mechanism.

Cylinder watches and carriage clocks with this escapement after cleaning, will often so increase in vibration that a banking error is introduced. A weaker mainspring is the remedy.

In a cylinder watch which has been cleaned a sticky banking will sometimes give trouble; brushing is often not sufficient to remove the holding tendency. Both pin and block should be scraped.

Isochronism.—The short arcs may be quickened by rectifying faults in the escapement.

By increasing the angle of draw on the pallets.

By altering the overcoil of the balance spring.

By altering the point of attachment of the spring to the collet.

By centrifugal force; *e.g.*—

A small hole is drilled through each of the half rims of the balance close to their fixed ends, and, after trial, broached out till the desired equality is obtained. This remedy was, I know, successfully adopted by the late B. Dennison. The greatest care must be taken to broach the holes alike, for if one half of the rim is weakened more than the other the balance will be out of poise when moving quickly. This method is open to the objection that the balance is rendered less rigid and more likely to be bent.

By using a shorter balance spring.

A shorter balance spring to give the same number of vibrations will, of course, be *thinner*, and to reach the hole in the stud will have more open coils.

If there is play between the curb pins; close them till the spring is but just free; there should be no apparent motion of the spring when its action is viewed through a glass.

By using a stronger mainspring.

As already explained (p. 63), the short arcs may be quickened by putting the balance out of poise, but this must be regarded as an illegitimate method, because the remedy is but ephemeral.

So far I have endeavoured to summarize the practice of various craftsmen. On request, Mr. W. N. Barber has favoured me with a method of procedure, which, from his high position as a theoretical and practical horologist, cannot fail to be a reliable guide.

"To find the size of the balance spring make a small prick in the board paper, put the balance pivot in this hole and make another prick against the end of a screw-head, the distance between the two pricks gives the diameter of the spring.

The balance being cut, trued and poised with the roller on in its permanent position, and each quarter screw turned out about one turn and a half, the spring may be selected, as described in pp. 49-50.

If no attention is going to be paid to the position of the internal point of attachment, and no interior terminal curve is to be used, break the eye of the spring out by putting it on a flat plate of metal and cutting the spring with a sharp graver, the eye must be broken out to such an extent that when pinned on the collet there will be a space between the collet and the first coil equal to the space between two coils of spring; hence, in breaking out, the bit of spring which will go in the collet hole must be allowed for. The pin having been fitted, as in p. 52, before pushing the pin home, set the spring as nearly flat as possible, judging of this by twisting the broach round in the fingers; then push the pin in tight by means of a very strong pair of tweezers, or small pliers; this must be a perfectly solid job or the spring cannot be got true and flat. When pushed in, the pin must not project at either end if the collet be of the usual round slit variety, because the coils of a close spring are so near that there would be danger of touching the pin end; if the collet be of the flat-sided type the small end of the pin should just peep through the collet, to facilitate a possibly needed pushing out.

If an interior terminal curve is to be used, break out the eye of the spring so as to leave a space equal to that which was occupied by about four coils; but an interior curve is not to be recommended since its practical disadvantages outweigh its theoretical attractions.

To get the spring true and flat, take a pivot-broach, blue it, file and stone round, and having a nearly straight part about ·05 mm. to ·1 mm. in diameter. Put this between the collet and spring, close up to the eye, and holding both broaches together in the left hand the spring can be bent inwards by pressing on it with a pair of tweezers beyond the broach; the broach being between the point where the spring emerges from the collet and the part of the spring being bent in; if it is required to bend the spring out, reverse the process, taking the broach a little farther along and pressing between eye and broach; the eye must be bent very carefully round by this or some similar method, and when approximately true and flat take the collet off the broach and put it on an arbor in the callipers or turns, a fine pointed pair of tweezers will now be required to complete the truing process; a very small eyed needle fitted to a handle and the end of the eye stoned off makes a useful tool for this purpose; if the collet be a flat-sided one, the spring can be easily got at, and can be bent by the aid of a very small pair of curved-jaw tweezers. Getting it true will probably send it out of flat, and so on, this must be got over by observing which way the spring goes out of flat when bent in for truth and then giving the spring such a twist out of flat in the opposite direction that the bending in for truth will bring it also flat. Unless the pivots are unusally fine there will be no danger in using the staff as the arbor to true the spring on. As a refinement it is desirable that the collet, with its pin and a little bit of spring in place, be itself in poise; with a flat-sided collet this will be nearly the case, but the slit in the round collet needs compensating for; this may be done by putting the slit at the back as usual and filing out with a seconds file, and polishing with a rotating rod of metal, a small hollow just in front of where the spring comes out of the collet; this hollow also allows the tweezers to get in to bend the spring. Poising the balance with the collet on is useless if this latter is not in poise and is going to be turned into a different position.

Next see that the stud hole is clear of shellac, soft bread, etc., and is broached out sound and parallel to the cock, then place the collet and spring on its seat on the balance, using a bone punch to press it down with, hold the spring with a pair of tweezers and count up the vibrations as explained at p. 49. The tweezers may have a brass slide put on as in a pair of sliding tongs, if thought advantageous. When to time, notice which screw-head, point or hole the tweezers holding the spring are opposite or nearest to, then put the spring between the curb-pins and pull through the stud hole until the above desired point is between the pins, the index being in the middle, then hold the whole affair up by the cock and count up exactly. By counting for two or three minutes as a final trial the average error should not exceed ·5 to 1 minute per day. When correct for time nip off the waste spring at the back, close up to the stud and remove balance and spring.

After seeing that the curb-pins are in good order, upright and just far enough apart to fit the spring, yet not grip it (if the pins require bending they must be bent with a knee bend, so that they are parallel to each other and do not form a taper slit), measure the radius of the spring and the radial distance from the centre of the cock hole to midway between the two curb-pins, multiply this latter by 100 and divide the result by the radius of the spring, the quotient will be the number of the terminal curve to be imitated. The measurement may be done with a slide-gauge having points.

The next step is to find how much spring will be employed in making this terminal curve, plus the little bit of spring used for elbow, plus the distance between the stud and curb-pins when the index is in the centre. If the length of the terminal curve itself is not known it must be found by bending a bit of waste spring up into a similar curve to the one going to be used, and then unbending this again into the radius of the exterior coil of the spring, add to this the two bits before mentioned and the point at which to start bending up the spring for the elbow is determined, Sets of copies of terminal curves can now be obtained, in which all the curves are made out of a half-turn of the exterior coil, this does away with all trouble in finding the length. The total length will be found to be generally about ¾ of a turn. **Next** to find the height to which the overcoil has to **be raised** above the plane of the spring. See **that the** spring collet is down **to** its place on the staff, place the balance and spring **in the watch, screw** the cock on, see **that the staff** end-shake is **about** right, **and then** look through the watch and notice what position the stud-hole occupies with **reference to the top of the** collet—perhaps **the middle of the hole may be level** with the top of **the** collet or the bottom of the hole level with the turning-off.

Next take the radius of the curb-pins, preferably with a small pair of beam compasses having a pump-centre, or a depthing tool; **put one centre** in cock-hole, adjust for upright and open until the other centre is midway between **the pins**; this distance may also be judged off by noticing **how many coils of** spring, counting from the outside, the **centre of the pins is** just over, and **afterwards** using this coil **as a guide or** template to **form the circular** arc of the overcoil along which the **curb-pins travel, by** looking squarely on top of the spring.

The next step is **to raise** the overcoil; **at this stage we** shall have to decide whether the bend upwards is to **be one** gradual slope or whether we shall have **two bends so as to** bring the overcoil parallel **to** the body **of the spring.** Both methods appear **to give about the same practical** results; **two bends require more** bending about of the spring than one, this **may be a disadvantage, but if a large model** Bréguet spring be examined it **will be at once evident that** the two bend variety gives a **much squarer motion of the** spring, and that the single long **slope causes an up and** down movement to the body of **the spring; on a small** scale, as in a watch, this is not, perhaps, **perceptible, but if** it exists we ought to **try and** avoid it.

Almost every springer has his **own way of raising the** overcoil, there is no **royal road, and however it is done much** will **depend on the judgment of the operator.**

Supposing **the single long slope is going to be** made, an easy way of doing so is as follows:—Put the spring on a **piece of brass or bone,** having a small depression in it, hold **the spring down when the overcoil is** to start with a pair of tweezers or with a bit of brass wire having a nick in its end to go over the coil at this point; now lift up the spring with a peg, putting the point of the peg under the spring in the hollow in the plate. Or it may be done simply with two pairs of tweezers, or in other ways. Many ways which are perfectly successful with large springs are often quite useless for small ones, or for open-coiled thin springs. The spring will now need twisting with two pairs of tweezers in about two places, one a little way up the elbow and the other, perhaps, half way to the end; the object is to bring the overcoil as nearly as possible parallel with the plane of the body of the spring, and also to raise it up exactly the right height for the stud. To judge of this latter hold the spring up sideways so that its plane is vertical, and see whether the stud end occupies that relative height with reference to the collet which was previously ascertained it should do. It is admissible, and, perhaps, advisable, even with the two-bend variety of overcoil, that the overcoil makes a very slight slope up to the stud, but on no account must it have an up and down shape, or the spring will wobble; if the stud hole is drilled very close to the brass work more slope must be allowed than would be otherwise desirable in order that the spring may free the cock when turned upside down. Making an overcoil with a knee-bend is not quite so simple, and a tool of some kind is almost absolutely necessary. A good and quick tool is shown in Fig. 39A.

It is made entirely of steel, with the jaws hardened, the jaws are about 3 mm. in breadth, and have four steel steady pins about ·5 mm. in diameter, fixed tight in one and a loose fit in the other jaw, these pins must be as far apart as the deepest spring to be used is broad, the jaw A is ground out with a radius of curvature about 4·5 mm., the jaw B with a radius of about 4·3 mm., this makes the jaws fit at their centre line and so grip the spring, while the slight divergence towards their edges allows the spring to be bent yet prevents it from buckling; the pins should be allowed to show at the back of A, they then form a guide to judge where the spring will bend. To use the tool put the spring in between the four pins at the place where the first bend is to be made; shut the tweezers together, and pull the spring up or down gently with a pair of roundish faced brass tweezers, or push it with a peg; to make the second bend pull the spring back through the jaws a little and make the spring butt against the opposite set of two pins. Now look at the back of A, and the end of a steady pin will show where the bend will come when made; if this would not be right height for the stud shift the spring either backwards or forwards a little until it is right, then bend down as before. These bends should be as round as possible and not abrupt angular ones. If a superlatively nice round curve is desired do not make the curve all at once, but bend a little and then shift the spring between the jaws backwards or forwards by a miscroscopic quantity and then finish the bending. Before taking the spring out of the tool it had better be bent to get the overcoil parallel with the body of the spring; to do this twist the overcoil up by aid of a pair of tweezers, taking hold of it close to the jaw.

Take the spring out and place it on some flat white surface, an enamel dial by preference, and look squarely down on to the spring and see whether the outside coil is bent in or out of place where it starts upwards for the elbow; if necessary correct this by means of a pair of tweezers and a peg.

The next step is to make the terminal curve itself, to do this place the spring on top of the photograph chosen to be copied, and gradually bend round to shape; get in the general outlines first and then form the exact copy, the part along which the curb pins are to work must be made to agree with the radius already measured with the beam compass, the pump-centre of this tool being placed in the collet-hole, adjusted for upright, and the other centre moved round just clear of the top of the overcoil. The curved-jaw tweezers used to effect this bending must not be of too sharp a curvature or the spring will show a series of distinct bends; the tweezers must previously have been carefully polished, especially along the highest portion of the convex jaw and the two edges of the concave jaw; these edges must be stoned nicely round and then polished by means of a piece of brass with a groove in it, and finally with a peg and diamantine; the tweezers should be held exactly upright, if inclined to one side they will bend the overcoil up and *vice versâ*, this property is often used to raise or lower a part of the overcoil.

Now prepare a flattened pin for the stud-hole in the same way as for the collet, and pin the spring in without the balance, the pin must be level with the front of the stud and just come through a little way at the back; set the spring about flat before pushing the pin home; now push the index along from end to end and see whether it moves the spring out of place, if it does it arises from the stud-hole not being drilled right for the pins; find out whether the pins are too near the centre or too far out by noticing which way the spring is moved, then push the index right to the fast, and with a pair of tweezers bend the spring either in or out with a knee-bend close up to the stud; finally the spring may be set quite square to the plane of the cock, and a look taken to see whether the cock-hole comes in the middle of the staff-hole through the collet, and also as regards the height of the overcoil.

Unscrew the stud, put the stud-screw up safe, put the balance in the watch and notice where, say, the edge of the stud will stand with reference to some screw in the balance rim when the watch is in beat; remove the balance and put the spring on in this position; oil the jewel-holes, screw the stud and spring on, and put the balance in its place in the watch.

The watch must be put exactly in beat by the drop of the escapement, noticing where the two drops occur and bisecting the interval between them. It will now be generally found that despite all care the spring is not exactly central, the deviation will, however, be very slight and may be corrected by pushing the overcoil in with a peg or pulling it out with a bit of fine wire filed very small, burnished, flattened and turned up at the end into a little hook, before doing this push the index right to the fast. The centrality of the spring is judged of by seeing whether the space between the two outside coils is the same all round. The body of the spring must, of course, be quite parallel to the cock, neither saucer-shaped nor like an umbrella; there must be no motion of the spring between the pins, the spring must not move when the index is pushed from end to end, the overcoil at at the elbow must not touch the second coil, and, finally, it must expand and contract concentrically; turn the balance round and so wind up the spring, and notice whether the coils touch on one side first, if not the spring should be fairly isochronal. If well sprung there will be one coil, or portion of a coil near the middle part of the spring which will neither move in or out when the watch is going, this is the "dead coil," and should always be looked for.

The watch should now be tried for mean time, and, if necessary, corrected by altering the timing-screws, taking a bit off two opposite screw-heads, putting in two heavier screws, or by fitting balance washers; there is no need, however, to get the watch exactly to time at this stage.

Two courses are now open to us—we may first adjust for compensation and then for isochronism and positions; or adjust for isochronism first, try the watch for positions to see if there be any serious structural faults, then adjust for temperature, and finally for positions. If the latter course be adopted the watch must be kept in a uniform temperature, during the other adjustments, if this cannot be done it is probably better to compensate first, since the isochronism should never be so much in error as to materially affect the compensation, then adjust for isochronism and positions, taking care only to alter in weight those balance screws which are quite close to the uncut end of the balance rim; if any material alteration is made either to the length of the spring (as by cutting the eye out), or to the weight of the balance it will only be prudent to try the compensation again, and, as the balance may take a fresh form by the heating, the positions will now again require trying. On the other hand the arrangement of the screws for compensation will affect the isochronism slightly, owing to centifugal force, but as the balance screws are put very nearly in their right places when the balance is made or cut this disadvantage is very small. For these reasons the general practice is to follow the second method. The watch may now be tried for isochronism as directed in p. 61, always taking the time at one place on the seconds dial of the watch. If the isochronism is defective it may be altered according to the following rules:—

To QUICKEN THE SHORT ARCS.—Looking at the point where the spring emerges from the collet notice where a radial line 90° from this point, counted along the spring the same way it is coiled up, cuts the terminal curve; bend the terminal curve IN towards the centre an equal amount on each side this imaginary line so that the line stands in the middle of the alteration in shape. Or if this alteration would come in an inconvenient place count 270° from the the point of attachment as before and bring the curve OUT towards the circumference along this line.

To SLOW DOWN THE SHORT ARCS reverse the above process.

These alterations will only need to be very slight indeed, and they must be effected with the curved-jaw tweezers as used for originally bending the overcoil, the spring being unpinned from the stud and placed exactly on top of the photograph from which it was copied.

THE POSITION OF THE INTERNAL POINT OF ATTACHMENT.—Both theory and practice show that the position of the point of attachment of the spring to its collet with relation to the pendant has a great effect on the position errors of the watch (especially if the spring has no interior terminal curve). If the displacement of the centre of gravity of the interior coils of a spring alone entered into the problem it would be easy to fix the most advantageous position of attachment, but when it is remembered that the staff pivots must have some shake in their holes, and that the lift of the escapement is not given as a pure couple, it will be evident that the virtual axis of vibration of the balance often does not coincide with its geometrical axis, and this in conjunction with the weight of the balance will cause the latter to have varying moments of inertia in different vertical positions; if to this cause of error we add the peculiarities of each escapement it is clear that the proper position of attachment cannot be predicated with certainty, and it is not to be wondered at that some timers doubt whether it is worth the trouble of putting it in the position indicated by the displacement effect alone, and so pin the spring on the collet just where it happens to come; but since we know in what direction the errors caused by the displacement tend, or in other words, we know the effect on the rates in the four vertical positions which are due to any given placing of the position of attachment, there is no reason why we should not avail ourselves of this knowledge, and make one error balance another by cutting out the eye of the spring so as to bring the point of attachment into the most suitable position for this particular watch, the selection of the position being made according to the observed rates of the watch itself. In a general way the position depends on the calliper of the watch, and it must be a matter of personal experiment to determine whether, for any particular make of watch, it pays to go to the initial trouble of getting the attachment into one particular place. It may be thought that the effect of the displacement of the centre of gravity of the spring can be got over by poising the balance and spring, considered as a system, in the watch according to the observed rates; this does not however by any means appear to meet the case, the explanation probably lying in some balancing effect produced by the pull of the spring on the collet and the lift of the escapement.

As regards poise what we really require is that the balance, with its rollers, collet and spring, shall be and remain in poise during the expansion and contraction of the spring.

It is safe to say that this is quite impossible. Even if the balance, etc., be perfectly in poise, and the spring is furnished with two theoretical terminal curves, so that the whole system is in poise when tried in the poising-tool, it will no longer be so when in the watch, because a certain proportion of the weight of the exterior coils is borne by the stud; much more will this be the case when there is only one curve. It follows from this that although the balance may have been poised in the tool it is not altogether illegetimate to alter the screws a moderate amount according to the rates shown, this is the universal way of correcting small positional errors, and is fully explained on p. 63.

Although this is the best practical way under the circumstances it is by no means satisfactory, for two reasons—the amount the balance spring is out of poise is a varying quantity as the spring opens and closes, and secondly, the virtual axis of the staff does not always coincide with its geometrical axis. If the positional errors are large, and faults in the machine likely to cause them cannot be detected or remedied, it is, therefore, better to cut out the eye of the spring according to the table on p. 33 or p. 35.

If it is decided to place the point of attachment in its most favourable position when first springing the watch the following method shown in Figs. 20 and 21 (p. 34) will be found advantageous, as it avoids a large spring eye. The spring being pinned on the collet as usual, we now require to know how many more degrees the overcoil will subtend when bent up than it does now when untouched. This is easily found, as before explained (a divided plate is here useful). Now place the collet central over o (Fig. 21), making the point of attachment lie on the line oc if the spring be a right-handed one, and on the line od if it be left-handed, and cut off the spring the determined number of degrees in front of the line oG, say along oM for example.

The weight of the balance must now be altered to bring the watch to time. In this operation it is often convenient to alter the weight of two opposite screws, and to be able to replace them, feeling certain that the poise of the balance has not been materially altered. This can be readily done by using a large, light compensation balance, all the screws except the quarter-screws having been removed. Notice which side each screw taken out of the balance of the watch belongs to, put them in the poising balance, adjust for poise in the poising-tool and remove metal at discretion."

Compensation Adjustment.

A hot and a cold chamber are usually required for the temperature adjustment. A stove or oven for high temperatures must certainly be provided. An ice box, essential when marine chronometers are being tested during the the summer, may, as a rule, be dispensed with for watches. A Norwegian refrigerator or box padded with a non-conductor can generally be so arranged in a cellar or other place of fairly uniform temperature so as to provide a receptacle with a sufficiently cold extreme, except in the very hottest weather. The adjustment for temperature is made after observations of the alteration of rate in the two extremes at which it is decided to expose the piece. A short exposure to the temperature, or a single observation, cannot be taken as a reliable indication of the effect, for the unnatural connection of the metals composing the rim of the balance requires time to settle.

The usual course is to place the piece to be tested, after its rate has been carefully noted, into the oven. After twenty-four hours, the rate is again noted. Say it has gained on its rate 8 seconds, It is then removed to the refrigerator, and subjected to the other extreme of temperature for 24 hours. At the end of that period a comparison shows that it has lost on its rate 7 seconds. Although the alteration in the two extremes is not equal, there is sufficient evidence that the balance is over-compensated. In the oven, the rims bending too far inwards reduced the effective diameter of the balance too much, and caused a consequent gain. In the refrigerator the rims expanded too much, and as a consequence the piece went slower. If it is a marine chronometer under trial, the weights have to be shifted a little towards the fixed ends of the rim. They must be shifted equally, or the balance will be thrown out of poise, and it is well to see that the slots in the weights are easy and do not grip the rim. If a watch is being tested, two opposite screws must be shifted towards the fixed end, care being taken not to screw them too tight to the rim.

F

To avoid mistakes impress on the memory the fact that a minus sign (—), that is losing on its rate in the oven, always involves shifting weights towards the FREE ends of the rim, and plus (+), or gaining, shifting weights towards the FIXED ends.

The piece is then again subjected to the extremes of temperature, and as the compensation adjustment gets closer the piece is taken from the refrigerator and placed a second time in the oven for verification before the alteration is made. As the trial proceeds the piece is allowed to remain more than 24 hours in each extreme, oftentimes a week.

Oven.—An oven for marine chronometers is generally a square box of sheet copper or stout zinc with one or two shelves. There is a jacket or casing round the top, bottom and three of the sides filled with water so as to keep the temperature uniform. The jacket may with advantage be covered with flannel or felt, except the bottom, which is slightly domed and heated by gas. The remaining side is formed into a door having double panes of glass let in so as to leave an air space between them. The oven is furnished with a self-registering maximum and minimum thermometer.

Hearson's Oven, shown in Fig. 40, is a good example tastefully arranged as a fitting for a watchmaker's shop. A is the water jacket; R and D plugs for filling and emptying; B gas tap; F shelves of coarsely woven wire on which the watches E are placed. The "thermostat" C is a very sensitive device on the principle of the aneroid, for keeping the temperature constant. It is a brass capsule, two inches square and quite flat, suspended from a tube running through the water jacket; this is filled with ether; when the ether boils the sides of the capsule bulge out, and by means of a rod (O), passing through the tube the gas is turned off at the tap. In this way a constant temperature of about 85° F. is maintained. There is a small bye pass for enough gas to keep the flame from being entirely extinguished. N is a thermometer. The water jacket though an excellent aid for ensuring an even temperature is not absolutely essential. The oven at Greenwich

Observatory is jacketted with glass wool, the waste from glass spinning, which is a good non-conductor. Many watch adjusters have for an oven but a wooden box with a sheet of metal under the bottom which is heated by a small lamp.

Mr. Schoof describes a cheap way of making an oven. To the middle of a thin iron plate (A Fig. 41) he fastens an iron tube (B), long enough to go right through a wooden box of the height the oven is to be, the iron plate being under the bottom of the box. The front of the box is formed into a door with double panes of glass as just described. A very small flame from a gas or oil lamp suffices to keep the highest temperature required for compensation adjustment. The one objection to the use of iron in the construction of an oven, is its liability, after hammering or rolling, to retain magnetism.

Fig. 40.

A very handy form of oven large enough for watches is shown in Fig. 42. There is an outer ring of thin sheet copper or zinc, 18 inches high and 12 inches diameter, with

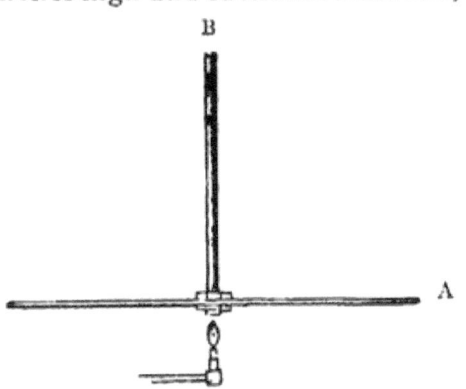

Fig. 41.

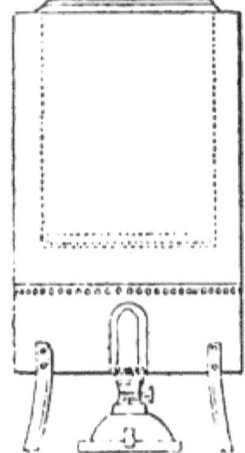

Fig. 42.

a horizontal partition 6 inches from the bottom, and with

three legs to raise it from the floor. Inside is a cylindrical copper or zinc box as shown by the dotted lines, 9 inches in diameter and 9 inches deep with a rim to prevent it going too far down, and a hinged bezil with double panes of glass with an air space between them. A disc of wood is placed on the bottom to rest the watches upon. The oven is heated by a paraffin lamp, and there are a number of holes in the outer ring just below the partition as shown. Round the sides above the holes are several layers of flannel, with a loose cover of the same for the top.

There is one advantage of a square oven over the cylindrical shape, and that is the facility with which the thermometer can be seen through the glazed door, but many good timers use a round one substantially as described, which is cheaper to construct.

Gas Governor.—Fig. 43 is Mr. Kullberg's effective and ingenious attachment to an oven heated by gas. The gas enters by the right-hand pipe into the glass tube, and from there through a hole in the left-hand pipe to the burner that heats the oven. It will be observed that the left-hand pipe is prolonged to nearly the bottom of the glass tube to furnish an attachment for a laminated arm. This arm, made of a thick piece of mainspring and rather thicker brass outside, carries at its free end a weight, and just beyond a conical tit that acts as a stopper for the hole in the exit pipe. A nick is made in this hole to allow of the passage of sufficient gas to keep the burner just alight even when the stopper is pressed home. As the temperature rises the bending of the laminated arm causes the tit to approach the hole, and the reverse action takes place with a fall of temperature. By tilting the instrument to the right or left the weight retards or assists the action of the laminated arm, so that any adjustment needed may be made. A screw with a large head nips the two pipes when the proper inclination has been obtained. There are also a pointer and circular scale for denoting the angle of inclination. The two pipes are of brass, but they may have flexible tube connections to the gas supply and burner respectively, to allow of the tilting of the governor, the lower part of which is, of course, immersed in the oven.

Fig. 44 is another form of a regulator, consisting of a glass tube with a reservoir of mercury (C) at the bottom. A hollow tapered glass plug is ground into the top of the tube. The gas enters this plug at A, and passes out of the bottom of the plug into the tube, and through B to the burner. The volume of gas allowed to pass depends upon the distance between the bottom of the plug and the top of the mercurial column. The screw allows the the temperature to be adjusted by enlarging or restricting the space for the mercury. A minute hole a in the side of the plug allows enough gas to flow to just keep the flame alight irrespective of regulation.

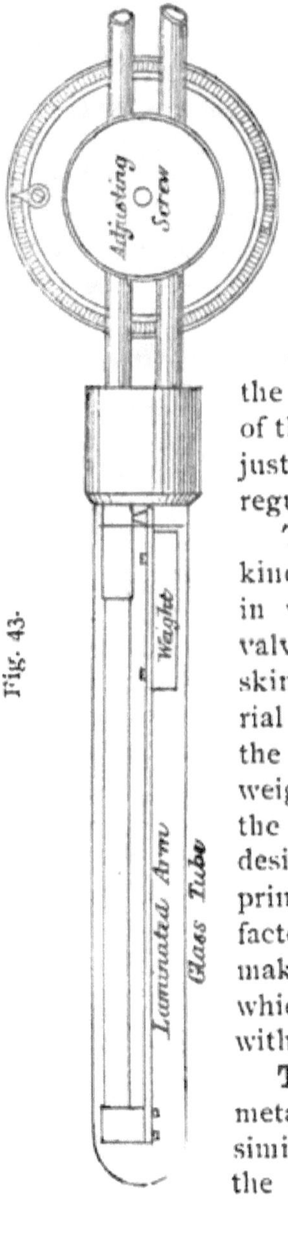

There is another kind of gas governor in which a sensitive valve of Persian sheepskin or similar material is interposed in the supply pipe, and weighted so as to keep the supply of gas at a desired pressure. This principle is not so satisfactory for watchmakers' ovens as those which deal directly with the temperature.

The Ice Box is a metal receptacle, very similar to the oven, for the chronometers or watches, surrounded

Fig. 43.

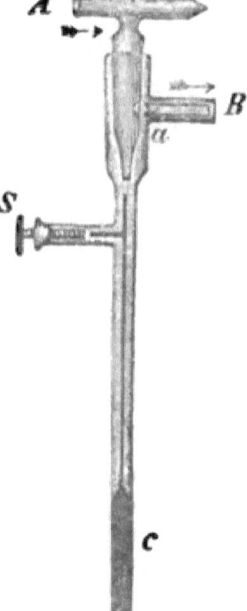

Fig. 44.

by a space for ice; the outer vessel has a tap to drain off the water, and is covered with some non-conducting substance.

Great caution is needed in using an ice box, to preserve the steel work from rust caused by condensation of the air. Mr. Arthur Webb places his chronometers in a brass box with a cover screwing against a waxed joint to exclude the air. Whenever this box is taken out of the ice box, it is allowed to remain unopened for about three hours, and as a further precaution the balance and spring are washed in benzine after a chronometer has been subjected to ice. If this is not done the chronometer, immediately after removal from the low temperature, should be placed for a period in a moderately warm receptacle, with an open vessel containing chloride of calcium, which absorbs moisture.

Balance out of truth.—To a balance maker setting a balance true is but a small matter. The following is related as the procedure of the late Mr. Earnshaw:—" He first took out the screws which he arranged in holes in his work board, and then cut the rim open. This caused it to spring outward, and he then placed it over a hollow filed in a piece of boxwood screwed in the vice, and rubbed the inside with an oval metal polisher. There were holes of different length and depth in the boxwood, and he also used polishers for rubbing according to the size and thickness of the rim. The first rubbing sent the rim in too much, and he pulled it outwards with his fingernail from the cutting. Of course the calipers were in constant requisition, and when the rim was quite circular, he subjected it to a heat test in order to 'set it.' This was done by putting the balance (with the staff) in an escapement box and holding it over the gas until the rim sensibly curved inwards. He remarked while doing this, that it was equal in effect to a six weeks' trial in the oven. After this he again set it true in the manner previously described, and then replaced the screws with due regard to proper compensation, at the same time getting the balance fairly in poise."

Mr. Plose has given the following excellent directions for truing a balance :—

"Before attempting to true a balance that is in bad condition, remove all the screws. Place them in holes made in the lid of a scape or pill box, in the same position as they occupy in the balance—blank holes to represent those in the the balance being also blank in the lid—they will then not become mixed. A rough sketch of the balance on the lid, with the holes made perpendicularly instead of horizontally, will be a help to identity. Be careful when unscrewing or screwing them in to hold the rim firmly in the fingers as near to the screw you are working at as possible, or it may bend, by this operation, out of shape.

Put the balance in the callipers, and test it for round and flat with a toucher of brass or German silver, screwed to one limb and having a long hole to allow of adjustment. Be sure the calliper sinks are not worn, or the job of making flat will become more difficult.

By looking down the edge of the toucher—which is best about the thickness of a sixpence, and perfectly flat and smooth for this job—on to the rim, you will be best able to see whether it is necessary to bend in or out. Before making any alteration examine your job closely, for a slight crack—almost imperceptible, mostly at a screw hole—is often a bar to a true balance. It is, in fact, almost safest not to do anything to such a thing.

Be prepared with brass-lined pliers, the noses tapering to about one-eighth of an inch, the linings hollow one nose and rounding the other, to fit the circular shape of the rim and to prevent it becoming marked. As these brass coverings cannot be filed to the right shape after they are on the insides of the noses, it is necessary to prepare them prior to fixing them. To do this, soften the pliers sufficiently to be able to drill them where the linings come, and steady pin a piece of brass, with a pin top and bottom, to each nose. The required shape can then be easily obtained by filing them when off, and testing their fit one to the other after being placed in position through the pins. Finally screw on with brass screws.

A gradual inclination to the staff, from the arm to the cut end of the rim, may sometimes be altered, by pulling this part gently out with the fingers; but any projections or indentations between these parts are best managed with the tool mentioned. A slight pressure, when the balance is gripped between the pliers at the spot where a sudden inclination takes place, and the rim from this spot goes in toward the staff, will bring the rim out to its proper place. The reverse may be obtained by holding the balance steadily and tightly with one hand, while the other uses the pliers to bend inward the part required. Take care, rather to do too little than too much bending in one operation. All alterations to be made with the balance out of the callipers. The toucher fixed by the screw in the position required, and a little red stuff applied to the rim, on the top, at the part to be corrected, will be a guide as to whether it has been bent enough, too much, or too little. After one part has been brought in, another close to it may require the reverse, or *vice versâ*. Should the balance be mounted out of round, it will be useless to attempt to true it by bending; but close examination will show if this is so, and a new staff that *fits* the hole in the balance is the best remedy. Many balances are spoilt through being badly mounted, becoming therefrom out of round or out of flat, sometimes both. The seat for the balance should be turned quite flat, otherwise there will be great uncertainty about riveting it flat; and the staff, as mentioned, should fit the hole, or true running will be out of the question.

The job of flatting may require an arm to be bent or bumped, or the rim to be bent or hammered. In most cases when the arm is bumped or bent, the rim will be found to have altered its position, but increasingly towards the cut end. This can generally be set right with the fingers, by holding the part at the root, near the arm, tightly between the thumb and finger of one hand while the other adjusts. A brass hammer or a brass punch will not deface the arm if it must be driven up or down with a blow, but the finger-nails often do the work as well. A sharp dent in the bar causing the balance to run out of flat is best removed with a brass punch like a blunt screw-driver at the face. Some smooth lead stakes of different sizes to go inside the rim, and one large one for any large size balance to rest on comfortably, with a hole for the staff, will be found very useful when it is necessary to recover its lost shape by impact. An ivory or boxwood hammer, 'loaded' at the opposite end to the pane—this part being used to make the rim flat when lying on the lead—and brass, ivory, and boxwood punches are very handy.

The flatting process mostly causes error in truth, and truing generally throws the balance out of flat; in consequence a repetition of the work—though to a less extent if care is taken—is often necessary. When satisfied with flatness and truth in the callipers, put the balance in the frame, without the roller, spin it gently round, and any little error still existing will show itself more perceptibly by watching a certain screw, or screw end, or anything close to the inside or outside of the rim as a guide to its being true. For the flat, observe the daylight above and below the balance when looking along the plate. I should have mentioned that, before anything is done that will raise or lower any part of the balance, it should be tried in the frame, and when running observe which part it will be best to operate on to also procure needful freedom; for time may be wasted in making flat in the wrong direction, when there is very little freedom above and below. The rim may be improved in appearance, if any slight scratches or marks are on it by polishing it with a boxwood slip and diamantine, the balance resting the while on a sound flat cork secured in the vice.

All being satisfactory, it only remains to replace the screws very carefully. Use a nice light screw-driver, with thin level blade, that will go well into the slit in the head."

Figs. 45 and 46 are two tools described by Mr. Robert Gardner. Fig. 45 is for truing the rim. The pin that is to be used on the brass side of the rim is of ivory, and the other of brass; the brass pin being used as a fulcrum for bending the rim in either direction by means of the ivory one. The ivory pin should be rather less than 1/16 inch diameter, and there should be a space of rather more than

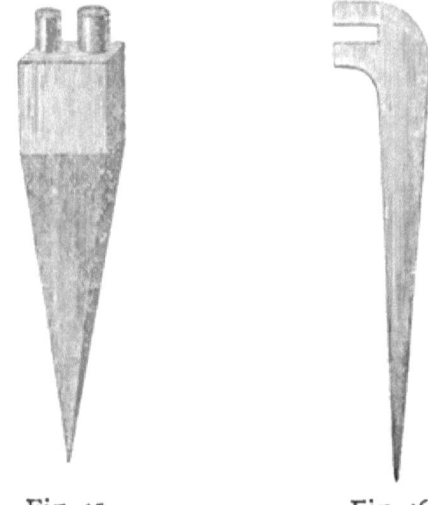

Fig. 45. Fig. 46.

twice the thickness of the rim between the pins. By means of this tool a balance can be very easily brought true in circle. To get a balance flat which has been distorted is sometimes troublesome, but it is greatly facilitated by using a piece of brass with a notch in it (Fig. 46) for slipping over the bar, so that the bar can be slightly twisted at the root without disturbing the position of the staff.

Poising the Balance.—To ascertain if a balance is correctly poised, or that the metal composing it is evenly balanced around its axis, the balace staff is centred between the points of callipers and one limb of the callipers gently tapped, when it the balance is not in poise the heaviest part is drawn to the lowest position. The sinks in the limbs must not be too large or worn. In Fig. 47 is shown an adjustable link connecting the upper and lower portions. By means of this, after the limbs have been opened to suit any particular balance staff, the link can be set so that if the callipers are opened to remove the balance they can afterwards be closed to clasp the staff again without fear of damaging the pivots, because the link stops them from going beyond their former position. As a poising tool for fine work, jewelled callipers are preferable. The ends of the limbs of an ordinary pair of brass callipers may be straightened as in Fig. 48, and a hole drilled and tapped in each; to these holes brass screws without slits in the heads are fitted; the

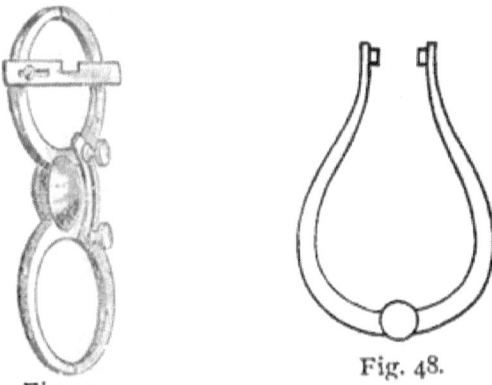

Fig. 47.   Fig. 48.

screw heads jewelled with good endstones and holes large enough to take any ordinary pivot. Some prefer V-shaped jewels instead of holes and endstones. One outside edge of the callipers is roughened, and when the tool is in use it is slightly rubbed with a screw-driver or other tool that may be handy, to cause the heavy part of the balance to travel to the lowest point. Plose's pattern is a very superior kind. It has ruby pins, with polished sinks and an adjusting screw to prevent the callipers closing too far and causing injury to the pivot ends from careless handling. The "Excelsior" callipers, having the limbs crossed as shown in Fig. 49, embody a distinct improvement over the ordinary kind. When held in the left hand, with the thumb and forefinger placed at opposite sides of the upper part, a slight pressure will close the callipers, while a pressure with the ball of the hand below the crossing of the limbs will open them as easily. That the old-fashioned callipers can only be opened or closed by the use of two hands is a disagreeable experience to every watchmaker, especially when the joint of the callipers works stiffly. Therefore this advantage in the "Excelsior."

The plyer form of callipers shown in Fig. 50 are excellent for poising. They are furnished with male and female centres and a spring for keeping the object to be trued in position without undue pressure.

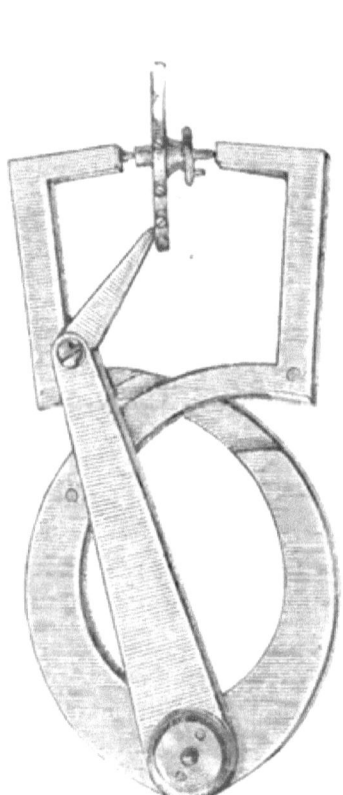

Fig. 49. "Excelsior" Callipers.

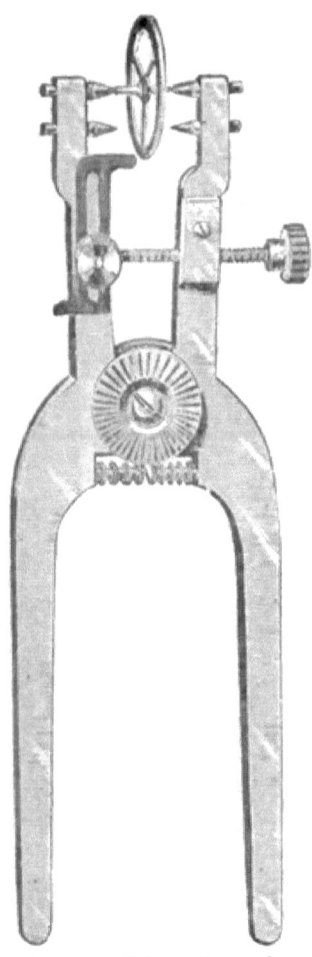

Fig. 50. Plyer-shaped Poising Callipers.

In the opinion of many, testing in the callipers does not afford a sufficiently exact indication of the truth of a compensation balance for fine work, and recourse is had to a special tool, of which the one in the drawing (Fig. 51) by Mr. R. Bridgman is a very fine example. The movable standard is kept in contact with the bed of the tool by springs, so as to be independent of the truth or otherwise of the traversing screw. The pivots of the balance rest on knife edges formed of sapphire, which may be adjusted to the desired distance by means of a screw. Screws on each

Fig. 51.

side of the standard serve to level the knife edges. Fig. 52,

Fig. 52.

the "Grossman," is a less costly instrument more generally used.

As to the correction of the poise, if the divergence is not great it may be adjusted by screwing in a timing screw or nut at the heaviest point, or by proportionately screwing in the adjacent ones. For larger errors it may be desirable to substitute a heavier screw of platinum for one at the lightest point, or to lighten a compensation screw at the heaviest point, or instead or in addition to interpose a thin collet or washer underneath the head of the screw at the lightest point. All these methods are adopted, and a choice will rest with the discretion of the operator. If the balance has been adjusted for compensation, screws approaching the free ends should be interfered with as little as possible in order that the compensation may not be unnecessarily deranged.

The Stud.—The stud for the attachment of the outer end of the spring has generally been a weak point. With the primitive English full plate stud screwed into the plate the spring had to be unpinned every time the balance was removed, and was as a consequence often distorted or broken at that point. In old verge watches it was nothing uncommon to find the spring attenuated and much shortened, for it was the practice of the repairer each time the end of the spring was broken off to correspondingly reduce the strength of the spring by rubbing it on an oil-stone. The more modern stud fastened to the cock by means of a screw is costly and not perfect. A careful workman before unscrewing the stud will move the index as far as possible to the fast position and then release the spring from the curb pins, but in unskilful hands the spring is often bent through disregarding this precautionary measure. Then, as a rule, the balance spring stud, when screwed to the plate or cock, admits of no alteration in its position, and though it presents the advantage that if removed it may be replaced with the certainty that it is unaltered with relation to the spring, it is undoubtedly a convenience to have a

Fig. 53.

stud which allows of adjustment, either in circle to bring the stud hole right for the spring, or to and from the centre of the balance, to suit the diameter of the spring. The adjustable stud shown in Fig. 53 answers the latter condition, and is found mostly in Swiss watches. The wing of the cock is formed into a slot, into which the body of the stud fits. The head of the stud rests on the top of the wing, and is kept in position by a cover plate and two screws.

Walker and Barber's stud, shown in sectional elevation (Fig. 54) and plan (Fig. 55) allows of adjustment in circle.

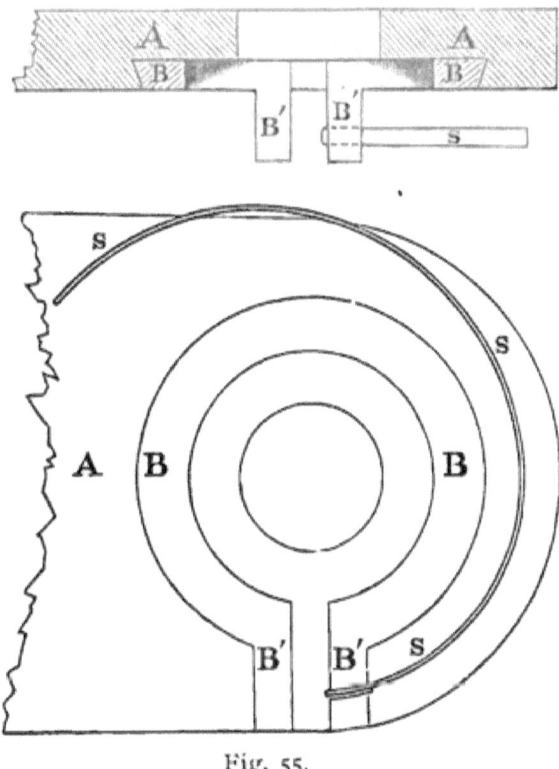

Fig. 54.

Fig. 55.

Fig. 1 = Longitudinal section through balance-cock.

Fig. 2 = Plan of balance-cock, inverted, showing stud and part of spring.

$A A$ = Balance-cock.

$B B$ = Steel ring sprung into balance-cock.

$B B$ = Projections from steel ring, one of which is drilled to take outer end of spring, the other to enable the ring to be contracted when it is desired to turn round or remove the stud.

$S S S$ = Terminal curve of spring.

Guye's stud also allows of adjustment in circle, Fig. 56 is a plan of the stud fixed under the balance-cock; and Fig. 57 a section through $a b$ of the stud attached to the cock.

In this case when the position of the stud is decided on

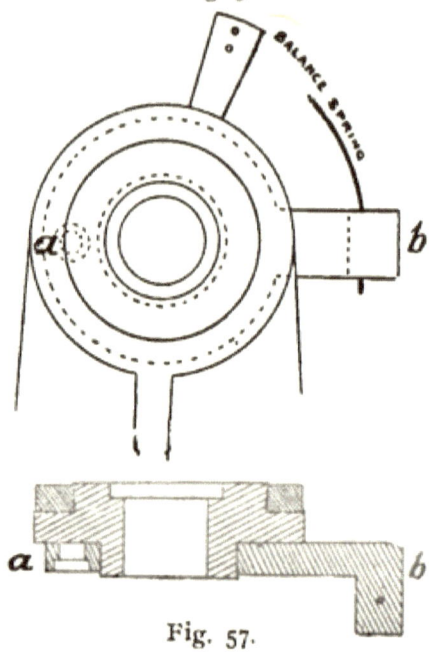

Fig. 56.

Fig. 57.

a screw at *a* serves to fix it to the cock.

Fig. 58 and 59 illustrate a stud described in *Deutsche Uhrmacher Zeitung*, and which appears to be especially suited for cheap watches.

As seen in Fig. 58 it is made of one piece of thin steel, the balance spring being clasped spring tight by the jaws, *b b*. The jaw *b* is a tongue punched from the body *k* and turned up *b'* being brought to it by forming the loop *h* and passing the end throught the slot *s*. The slot *s* from which *b* is cut serves for the screw *e* to attach the stud to the cock *B*, as seen in the sectional view Fig. 59. *S* is the balance spring; *R* the collet, and *z* the regulator.

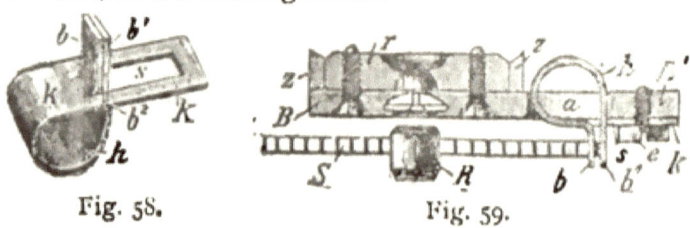

Fig. 58.   Fig. 59.

This stud allows the spring to be clasped at any height to suit the collet, and also permits adjustment to the diameter of the spring. By its use the buckling of the spring by pinning into a round hole is avoided.

Stud Remover.—Mr. Daldorph has devised a useful table with four slits as shown in the sketch (Fig. 60) for

Fig. 60.

supporting a Swiss cock while pushing out the stud. The holes in the table, and also the slits, allow the tool to be used as a support when pressing a balance spring collet to position.

The engraving (Fig. 61) shows a pair of tweezers for removing studs such as are pivoted to the balance cock.

Fig. 61.

By placing the tweezers so that the slit embraces the stud and the projecting finger is pressed on the pivot, a tight-fitting stud may be readily removed without fear of damaging the balance spring.

CHAPTER V.

The Manufacture of Balance Springs.

So long as the verge escapement was in use watch balance springs were usually coiled from soft or hard-drawn wire with the aid of curved nosed tweezers, or a coiling pin, which latter was very often simply a round broach. The manner in which the coiling pin was used to bend to a

volute the wire as it lay on the thumb of the operator will
be understood from the sketch (Fig. 62). Starting with
the eye of the spring each coil was finished before the next
was begun, and many of the springs so made were marvellously
true. Inaccuracies in flatness were remedied in
bluing. For this purpose the spring was placed in the
centre of a copper pan and held down by a skeleton disc
on the free end of a springy arm, the other extremity of
which was attached to the handle of the pan.

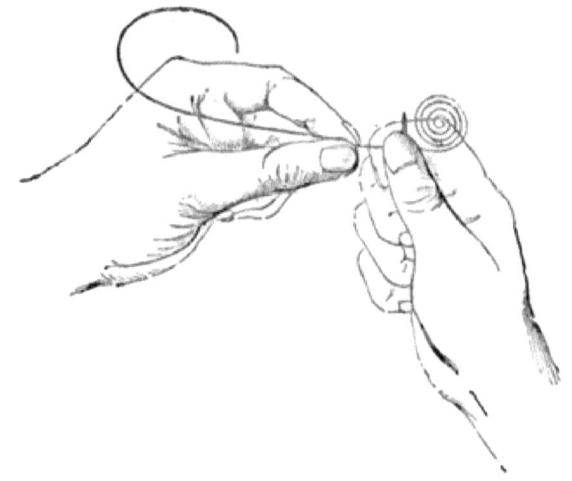

Fig. 62.

Soft balance springs soon get out of shape and are therefore
not reliable, but their strength is not materially
increased by hardening. If a watch is brought to time
with a soft spring, and the spring is afterwards hardened,
the rate of the watch would be accelerated but a few seconds
in twenty-four hours.

For many years certain Swiss spring manufacturers have
prepared springs of very hard drawn wire, which, being
comparatively inexpensive, serve a useful purpose. They
are much more lasting than soft springs, but more brittle,
and not so good as the best fire hardened springs. At one
time springs made from hard wire were supposed to be
hardened after being made by some chemical process, and
the secret of their manufacture was well kept.

Arnold invented the helical spring for chronometers, and

patented it in 1776. To form the spring he wound the wire on a spirally grooved cylindrical block and hardened it, practically just as such springs were prepared till quite recently.

From this time springs for cylinder and duplex watches though turned up by hand were sometimes hardened.

The method of hardening a flat spring without distorting it after it had been applied was at one time kept a secret by the few who practised it. It is however exceedingly simple. The spring is placed on a round, PERFECTLY FLAT plate of German silver; the space between the coils being filled in with black lead, the spring is covered with a plate similar to the first. These plates, with the spring between them, are then laid on a small press of German silver very much resembling in form the presses used for copying letters. (See Fig. 63.) The plates are kept together by

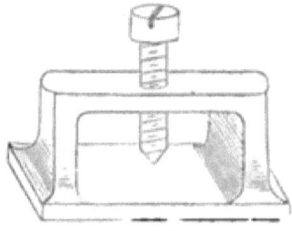

Fig. 63.

means of a screw. The spring, press and all, are then heated to a cherry red and plunged into water. When cool a thin slip of bright steel is laid on the press, which is heated till the slip of steel is brought to a blue colour, when it is plunged into oil. It is not imperative that the press should be of German silver, but it must not be of steel or wrought iron, or the springs will be spoiled. A press of this sort is also useful for restoring to shape a spring distorted from flatness. This may be placed between the flat plates and heated to a blue, as already described.

Modern Hardened and Tempered Springs.— In making balance springs the first requirement is good wire. After being drawn to suitable sizes, the round wire is flattened between rollers and then drawn through jewelled claws or rollers to ensure uniformity of substance. The most general fault is that, during the frequent annealing necessary in drawing it so small, the steel has lost too much of its carbon, and is difficult to harden except at an excessive temperature. Before making balance springs it would be prudent to test a piece of the wire for hardening and tempering. If the wire is suitable it should harden well after heating to what is called a cherry red, that is before it attains a brilliant red hue. Flat spiral springs are coiled up in a circular box, like a small watch barrel, and cover (see Fig. 64), made preferably of aluminium-bronze or of platinum. Copper and German silver are sometimes used, but platinum and aluminium-bronze are found to best retain their form after heating. Obviously the top and the bottom of the box must be perfectly flat if flat springs are to be produced. Though rigid the box and cover must be as thin as possible, because if a considerable body of metal surrounds the spring it has to be raised to a greater heat to harden. Either two, three or four springs are coiled at a time, generally three; if coiled up four at a time the coils would be more open than usual; springs for overcoils are generally closer and would be coiled two at once.

The following is substantially the process described in the *Horological Journal* by Mr. Glasgow:—" The winder

Fig. 64.   Fig. 65.

(Fig. 65) is of steel, with a pivot formed on the end to pass through a hole in the bottom of the box, and project into the box just the width of the wire of which the springs are to be made. Across the pivot of the winder are slits to receive the ends of the wire. For springs, coiled up two together, there is one straight slit across the centre of the pivot. If three springs are to be coiled up together, three equidistant radial slits are cut (as at *a*, Fig. 65), and for open springs coiled up four together, two slits at right angles across the centre are made. The pivot should be snailed from the slits according to the wire to be used, so as to ensure the truth of the eyes of the springs. A hole is drilled and tapped in the centre of the pivot to receive a small screw, the head of which passes through a hole in the cover. Holes not radial, but almost tangential, are drilled through the side of the box to thread the wire through. The ends of the wire passed into the slits are secured by the screw, the cover is gently pressed on, and the winder rotated till the box is full. The ends of the wire are cut off, the screw and winder removed, and the cover bound tightly with wire. The holes are then stopped, usually with a mixture of soap and animal charcoal to exclude the air, and the box put in an open iron vessel containing charcoal, which has been placed in the fire of an ordinary grate." Animal charcoal is usually prepared from burnt bones or burnt leather reduced to powder. The object of using it in this connection is to further enrich the steel, but some discretion is necessary, for if the steel is already sufficiently rich in carbon it may be rendered too brittle. Difficulty would then be experienced, especially in bending overcoils, if such were needed. If the steel is found to harden readily at a proper temperature vegetable charcoal may be used to exclude the air. If soap is used it should not be wetted, but softened by heating. Either beeswax or oil appears to be a more suitable vehicle than soap. When the spring box is cherry red, it is dashed into a vessel containing plenty of cold water. The spring box is then placed in a boiling out pan or an old metal spoon filled with oil, and held over a flame till the flame just flickers over the oil.

After cooling, the springs may be turned out of the box. If they do not come apart quite readily they should not be forced, as they would be likely to be permanently distorted thereby. If thrown into a pill box and rattled for a minute the slight concussion against the sides of the box is usually enough to induce separation unless the springs have been burnt. Water or oil is the medium generally selected for plunging steel in to cool it when it has been heated to the requisite redness for hardening. Either mercury or petroleum may be recommended if extra hardness is desired. Salt water will give great hardness, but the steel is rendered brittle. Oil and mercury are considered the two best media for hardening steel if toughness is desired. If the steel has been protected from the air, and not overheated during the process of hardening, its surface will not be scaled nor materially injured, and its brightness may be restored by polishing, but if the surface is intricate, as is the case with balance springs, a quicker method of cleaning may be employed. The steel may be washed in a strong solution of hydrochloric acid (about one-third of pure acid to two-thirds water is recommended), and immediately afterwards rinsed in a saturated solution of cyanide of potassium. In one respect springs finished with a high polish are inferior to others left grey, for they are more liable to rust.

To avoid oxidation of the surface during hardening and tempering the springs melted crystals of potassium cyanide may with advantage be employed as a heating medium. The cyanide having been placed in a wrought iron pot on a stove and raised to a red heat, the spring box is immersed and allowed to remain till it becomes of the temperature of the bath, when it is removed and plunged into cold water. A film of cyanide clings to the exposed surface when the box is removed from the bath, and so the air is entirely excluded from contact with any part of the springs. This protecting film leaves the surface when the box is plunged in the water.

To temper the springs, the box on being removed from the water may be again momentarily plunged in the cyanide bath and then subjected to an even heat in a small stove or iron box. When the film of cyanide peels off the box it is considered the desired temper is obtained.

This is, I believe, the process adopted by some of the Swiss spring manufacturers, and in America. Of course, wire of uniform quality must be selected to be sure of always attaining the desired temper. The difficulty generally is to ensure uniformity of quality, and for this reason no exact tempering temperature for balance springs can be given, because the steel varies in quality. Continual care is necessary to ensure success. Mr. Arthur Webb, on taking a fresh bobbin of wire, hardens a piece, and then finds the temperature at which that particular wire yields the desired temper, by placing it in the oil bath with a thermometer. When satisfied with the temper he notes the temperature which he assumes will be the best for tempering so long as that wire lasts.

Still, the following table, giving the characteristics of a wide range of temperature may be useful :—

Colour.	Purpose.	Temperature.	Alloy, whose fusing point is of the same temperature.	Effect on Tallow.
Pale straw	Lancets and Tools for Cutting Iron	420° Fah.	7 lead 4 tin	Vapourizes.
Straw	Watchmakers' Tools	450° ,,	8 ,, 4 ,,	Smokes.
Straw yellow	Pen Knives and Razors	480° ,,	8½ ,, 4 ,,	More smoke.
Nut brown	Small Pinions and Arbors	500° ,,	14 ,, 4 ,,	Dense smoke.
Purple	Large Pinions and Arbors	530° ,,	19 ,, 4 ,,	Black smoke.
Bright blue	Swords and Watch Springs	580° ,,	48 ,, 4 ,,	Flashes if light is applied.
Deep blue	Watch Balance Springs	490° ,,	50 ,, 2 ,,	Continuous burning.
Blackish blue	Chronometer Balance Springs	640° ,,	All lead or boiling linseed oil	All burns away.

Steel is less oxidized by tempering in an alloy than if tempered in the air, and the required temperature is obtained with much greater certainty. Olive oil, which boils at 600°, is now most generally used as a bath for tempering balance springs. But great care must be taken to ensure the purity of the oil. Olive oil adulterated with cotton seed oil, when used as a tempering bath, has been found to rust the springs, possibly from the presence of some chemical employed in the preparation of the oil. Pure olive oil may be used with confidence. It may be heated in a wrought iron pot, which must not be tinned. The cover may have a hole in it for the insertion of a thermometer, and hooks on the under side from which the springs may be hung.

Springs are more liable to distortion when perfectly hard than when tempered. No explanation appears to have been given of this curious fact, but, mention of it may prevent a beginner attempting to polish watch springs before tempering.

The most usual method of mechanical polishing is as follows:—

The flat faces are polished by gently pressing the tip of the finger on the spring, and moving it in a circular direction on a piece of writing paper on which red stuff has been rubbed.

The spring is drawn down over a piece of wood with a conical end like Fig. 66, with a pin at the apex for the eye to go over, and the outsides are polished with a well-worn brush charged with red stuff.

The inner sides are much more difficult to polish; but if the wire is good, and has not been raised to an excessive heat in hardening, and if the air has been properly excluded during that operation, but little polish will be needed. The spring is placed on a flat piece of cork, and the inner sides rubbed to and fro and from side to side with a finely-

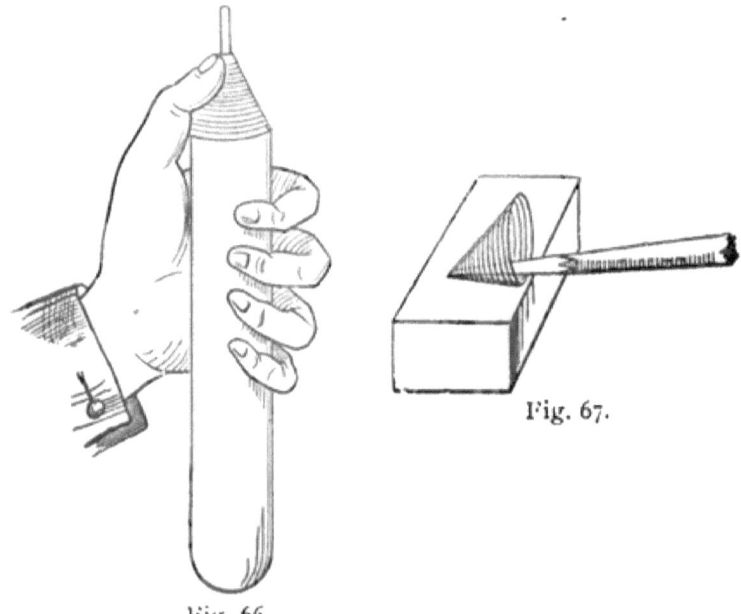

Fig. 66.   Fig. 67.

pointed peg charged with red stuff, the spring taking the form of a cone the while, as shown in Fig. 67.

After polishing, the spring is washed in PERFECTLY CLEAN benzine, and blued on a flat plate. The plate should be heated before the spring is brought into contact with it and care taken that no dust is allowed to settle on the spring. A glass tube may be placed over it during the operation.

An alternative method of polishing flat springs on the surface and between the coils at the same time is to place the spring on an india-rubber pad under two plates with radial sides terminating in a point, as in the subjoined sketch, Fig. 68. Over the spring is a holder on a universal

Fig. 68.

joint. Through a hole in the holder passes freely the polisher spindle, carrying at its lower end a thin slip of willow wood cut almost to a knife edge where it bears upon the spring. A little pressure on the knob at the top of the polisher holder will cause the coils of the spring to make serrations in the edge of the willow. Charged with a little fine emery the polisher is now pressed on the spring and rotated as far as the radial edges of the plates will allow for, say, six times backward and forward; the position of the spring is then altered so as to bring the part at first under the plates into contact with the polisher and treated in like manner. The spring is then turned over and the other side operated on. The india-rubber pad is on a circular metal table, and underneath the tool is a bridge; one end a helical spring around the stalk of the table presses against the underneath surface of the table and the other end against the bridge, and so keeps the spring on the rubber pad sufficiently tight against the plates. Care must be taken not to rub the spring more than is necessary and to subject each part of the surface to the same number of rubs. Excessive and indiscriminate rubbing will spoil the spring.

Fig. 69.

Fig. 69 shows a different arrangement for winding springs, for which I am indebted to Mr. R. B. North. Here the winding machine stands on the bench, and the centre arbor is actuated by turning a handle at the side with which it is connected by a pair of mitre wheels. The spring box without the cover is placed on the table of the machine and prevented from turning by a notch at the edge which engages with a pin in the table. At the end of a hinged arm is pivotted a little bar which may be pressed on the wire while it is being wound in. In this way the winding is under perfect control, and can be watched as it proceeds.

When the box is full the cover, with a hole to pass over the centre arbor, is placed in position and fastened with three screws. There is a special feature in the box which is worthy of mention. It has a loose disc or false bottom which can be readily trued or a fresh one substituted. The cover is flat, and does not enter the box, therefore narrow wire would not fill the space between the disc and the cover, but the central hole in the bottom of the box is larger than the hole in the disc and tapped, so that when the cover is screwed on and the centre arbor removed a screw can be inserted to press the disc till the whole surface of the wire is in contact with both the disc and the cover.

J. F. Cole used instead of a spring box a square plate, rising from which were four half round pins instead of a rim, keeping the wire within the required limit. Helical springs are formed by winding the wire on a thin solid drawn tube of brass or aluminium-bronze. If the same tube is to be used for tempering it has shallow grooves cut to the pitch the spring is to be, and in depth about half the thickness of the wire. The ends of the spring are fastened to the tube under the heads of brass screws which must be made left-handed, so as to draw the wire tighter on the tube as they are screwed up. The tube is mounted on an arbor in the turns; one end of the wire is fixed by one of the screws, the remainder of the wire hangs freely in the air, a weight of about 12 ounces being fastened to the lower end to keep the wire taut. A piece of thin sheet platinum or sheet copper is wrapped round the spring and tube and kept close with binding wire prior to hardening. It is hardened by being heated to redness and plunged into cold water. Immediately it is withdrawn from the water one of the screws is slackened, the spring drawn tight on the tube and held again in position by the screw. It is then tempered in boiling oil.

A grooved tube is not essential for coiling the spring. Mr. Mercer uses a plain thin tube of solid drawn brass or aluminium-bronze. The wire is coiled closely, covered with a thin sheet copper wrapper on the inner side of which has been smeared either beeswax or a little lard; the ends of the copper are tucked over the ends of the tube which is bound round with wire. A light wire holder, sufficiently long to protect the operator from the heat, is attached, and the tube is subjected to a charcoal fire heated by gas. The operator turns the tube, watching till it attains a cherry red, when it is dashed into mercury, which Mr. Mercer prefers as a hardening medium. During hot weather the pot containing mercury should be surrounded with cold water or the steel may not harden satisfactorily. After hardening, the spring is tempered on a grooved tube. The lard should be melted fresh pork. Lard, as purchased, is liable to contain salt or other impurity which might engender rust.

The outsides and edges are polished on a block a little larger than the tube used for coiling the springs. For polishing the inside, the spring is mounted on a piece of wood charged with red stuff and rotated in the turns, the spring being held between the thumb and finger the while. Before bluing the spring is thoroughly cleaned with soap and water, and afterwards with benzine. Unless the spring is perfectly clean it will not blue evenly. For bluing, the springs are secured by screws to a block with very shallow grooves very similar to the tube used for tempering, except that it is solid and a shade larger in diameter. The block is placed end upwards on a bluing pan, which is heated over a spirit lamp. The spring may be encircled with an open glass tube to keep the temperature uniform, and keep off dust which might make the spring specky.

Mr. T. Hewitt's procedure in coiling and hardening is a little different. He turns down his piece of solid drawn brass tubing, leaving the surface plain without grooves. On this he winds the wire, the coils following each other close together. He smears a mixture of oil and vegetable charcoal on the coiled wire, and over all slips another thin brass tube, stopping the ends of the two with a luting of the oil and charcoal. It is then hardened. After hardening, the spring is wound on to another brass tube having shallow grooves of the desired pitch and of exactly the diameter at the bottom of the grooves as the hardening tube was, fixed with left-handed screws and tempered. Tempering sets the spring satisfactorily to the pitch decided on.

The late F. Knudsen, a very successful chronometer springer, wound his springs on a plain tube, and heated them for hardening by hanging the tube on a pin in a piece of charcoal and directing a flame through the tube with a blow-pipe. After plunging in water he heated the tube to nearly the ultimate tempering heat to avoid breaking the spring in loosening the screw. He set the springs by tempering them on a grooved tube, and after polishing finally blued them on a plain block, the spring being secured at one end by a screw.

After helical springs are tempered the greatest care is taken to avoid rust and stain. I have heard of springers conducting subsequent operations necessitating handling with stalls of oiled silk on their fingers to ensure that when the terminals of a spring are formed, and it is primed on, the even colour remains without blemish. This, though, is not the usual practice, and is, I should say, not necessary unless the operator has very damp hands. Indeed, rust is generally the result of dirt or impurity met with in course of manufacture.

CHAPTER VI.
Non-Magnetic Material and Material Insensible to Changes of Temperature.

Watches with quickly moving parts of steel are rendered unreliable if exposed to the influence of magnetic or electric currents, and as the number of electric appliances for lighting, traction, and other purposes is continually increasing it appears to be evident that some other material than steel must be found for such parts of watches as the balance, balance spring, and escapement.

Aluminium-bronze, which combines strength with lightness, is particularly suited for the lever and pallets. The steel balance staff, pallet staff, and escape pinion may be retained, their circumferential velocity being small. A steel seconds hand must not be used, and steel hands had better be altogether avoided. For the ordinary run of watches, a plain gold or brass balance may be used. Many attempts have been made to devise a compensation balance, in which the use of steel is dispensed with.

Frederic Houriet, in the early part of the century, produced a non-magnetic watch using a compensation balance of platinum and gold. J. G. Ulrich took out a patent in 1828, and two subsequently, for non-magnetic balances. His claim seems not to have been based on exhaustive experiment, though it embraces nearly every possible metal. Messrs. Arnold and Dent made many experiments between 1830 and 1840, and they used balance springs of glass, of gold, and of palladium. Briefly, their greatest success with compensation balances was obtained with a construction of platinum and silver, which compensated fairly well, but were lacking in rigidity.

C. A. Paillard, in some instances, appears to have used a palladium alloy for the inner part and brass for the outer part of the rim, and in others to have formed both laminæ of different alloys of palladium. Professor Houston, in an exhaustive paper to the Franklin Institute, gave the following interesting particulars respecting Paillard's alloys:—

"The alloy described in the first U.S. specification contains the following metals, viz.:—

> Palladium .... 60 to 75 parts.
> Copper.................. 15 „ 25 „
> Iron .... 1 „ 5 „

"We will for convenience call this alloy No. 1.

"The specification states that the preceding proportions, or percentage may be somewhat varied, without appreciably affecting the essential characteristics or properties of the alloy.

"The process for the production of this alloy is as follows: About half of the palladium to be used is placed with the other metallic ingredients, and with a small quantity of borax and powdered charcoal, in a clay crucible and heated until melted. The remaining part of the palladium is then added, and when the whole is melted the molten mass is poured into a suitable mould, and when cooled is ready for use.

"U.S. Patent, No. 367,159, describes an alloy cheaper in its production and suitable for a lower grade of watches, or for certain parts of watches.

"The inventor makes the following statement in the specification: 'I have found by experiment that the alloy hereinafter described and claimed can be successfully employed in the manufacture of watches and timepieces for all the parts not required to be non-magnetic and hardly dilatable, as it is not perceptibly affected by ordinary magnetic or other disturbing causes or influences. The cost of its production is also such that it can be profitably and economically used for all the ordinary parts of watches and for the mechanism of the cheaper grade of watches, where perfect uniformity and regularity of movement under all circumstances is not required.'

"The composition of the above alloy (No. 2) is as follows:—

Palladium	50 to 75	parts.
Copper	20 ,, 30	,,
Iron	5 ,, 20	,,

"The method adopted for its manufacture is the same as that described for the first alloy.

"U.S. Patent, No. 367,160, describes an alloy possessing in the highest degree the properties desired in parts of high-grade watches liable to change of rate.

"The following is the composition of the alloy (No. 3) that is described and claimed in this patent:—

Palladium	65	to 75	parts.
Copper	15	,, 25	,,
Nickel	1	,, 5	,,
Gold	1	,, $2\frac{1}{2}$	,,
Platinum	$\frac{1}{2}$	,, 2	,,
Silver	3	,, 10	,,
Steel	1	,, 5	,,

"The same method is used in the manufacture of this alloy as in the preceding alloys.

"U.S. Patent, No. 367,161, describes an alloy which, besides possessing the properties of the preceding, possesses the additional property of being raised by tempering to a very high degree of hardness. He uses the alloy as follows: 'In order that the best result may be obtained in watches and chronometers, I have found that it is necessary that other parts of the mechanism employed therein, viz., the escape-wheel, escape-lever, guard-pin, and regulator-index shall possess the same characteristics and properties, and in addition thereto be capable of being tempered to a high degree of hardness to prevent wear and abrasion.'

"This alloy (No. 4) has the following composition, viz.:—

Palladium	45 to 50	parts.
Silver	20 ,, 25	,,
Copper	15 ,, 25	,,
Gold	2 ,, 5	,,
Platinum	2 ,, 5	,,
Nickel	2 ,, 5	,,
Steel	2 ,, 5	,,

"The process of manufacture of this alloy is the same as that used in the other alloys.

"The following experiments were made to test the non-magnetic character of the palladium alloys and the extent of the protection afforded to the watches before alluded to, when placed in various magnetic fields.

"Two palladium balance springs of different alloys were placed in a uniform field, the direction of the lines of force in which was determined by a very thin layer of iron filings. On placing the springs in various positions in the field no change in the grouping of the lines was observable. They were neither concentrated on the palladium alloys nor repelled from them; that is to say, they were neither appreciably paramagnetic nor diamagnetic.

According to the experiments of Faraday and the later investigations of Plücker, palladium is a paramagnetic substance; that is to say, it concentrates the lines of force upon it after the manner of iron. Its paramagnetic properties are, however, comparatively feeble.

"Bearing in mind the paramagnetic character of many of the components of the preceding alloys, their failure to exhibit any of the properties ordinarily recognized as magnetic is interesting from a scientific standpoint. Thus in alloys Nos. 1 and 2 the copper is the only component that is diamagnetic. In alloy No. 2, the steel and nickel are powerfully paramagnetic, and the palladium and platinum are also paramagnetic. The copper, gold, and silver only are diamagnetic. In alloy No. 4, all the components are paramagnetic except the copper, silver, and gold.

"The masking of the paramagnetic properties of some of the components of an alloy would seem to point to the probability of such alloys being formed by true atomic or chemical combination.

"In order to observe the effect of an intense field on the springs they were placed in the very powerful field of an electro-magnet, the massive pole-pieces of which were but a few inches apart. No deflection of the light springs was observed when suspended in this field. When allowed to fall through the narrow gap between the pole-pieces, they fell quite freely. The intense field failed to produce any appreciable magnetism in them.

"Similar experiments were tried with the compensating balance with the same result."

Professor Houston says the segments of Paillard's balances consist of two laminæ of different palladium alloys, the co-efficients of expansion of which are so proportioned as to permit them to act as in the compound balance of brass and steel. This does not agree with the prescription in Paillard's English patent, which clearly defines the inner part of the rim as palladium alloy and the outer part as silver or brass. It adds :—

"After melting the alloy, the molten metal is poured into a mould to form a plate. This plate, when cold, is forged or hammered, being frequently heated to red heat, if becoming too hard, until the desired thickness is obtained.

H

There is then stamped out, by means of a die, a disc, in the centre of which a hole is pierced, and the disc is turned true. The hole and the one side of the disc are then filled up and covered with a fire-proof clay, or any such suitable material (plumbago, for instance), which is allowed to dry on the disc, so that the latter, in all its parts so covered, will not be touched by the second metal in the subsequent operation. The so prepared disc is now inserted in a crucible, in which pieces of brass or silver and borax are placed and melted, the crucible being introduced into a muffle furnace. The molten metal will form a rim around the disc. If preferred, the silver rim may be produced by depositing this metal by a galvanic process.

"The melting of the metal around the alloy disc will require some skill or experience, inasmuch as care must be taken that the metal is not heated more than required for melting the same, so as to avoid melting of the alloy disc, or even to avoid the molten metal attacking the alloy disc. If the rim or ring is made of brass, the melting of the latter may be facilitated by an ample addition of zinc, which may form 40 to 50 per cent. of the brass. In using silver, about 20 per cent. of zinc may be added. The ample addition of zinc offers this advantage, that the metal becomes more expansible, and thereby increases the function of the bimetallic balance. If silver is deposited by the galvanic process, this will require the employment of screws or heavy masses, for the compensation to be a sufficient one.

"The rest of the work for making the balance is entirely mechanical, and does not differ materially from the method heretofore in use.

"First, the excess of metal is removed by means of a file or a hand-tool lathe, and then the centre part of the disc, either from one side, or from both sides, is hollowed out, so that in this centre part there remains a thickness only, corresponding to the thickness of the middle arm, which is to be cut out, care being taken to preserve a bimetallic rim of such width, that it is composed by $\frac{1}{3}$ of palladium alloy and by $\frac{2}{3}$ of brass and silver. By partially removing the remaining central part of the disc (by filing or otherwise), the central arm or arms for the central pin are formed, so as to complete a balance.

"The boring of the screw holes and the cutting of the thread in them, is made by means of a dividing engine.

"The arm or arms of the balance, as well as the interior of the bimetallic rim are generally tempered, whereas the exterior, and the faces of the said rim, may be polished.

"It is desirable, in order to give hardness, to heat the balance in the same way as for tempering balance springs. The bimetallic rim of the balance is then cut towards the arms, and provided with its screws or set weights."

M. Rambal, describes a balance with the inside of pure malleable nickel and the outside of brass : good timing was obtained, but the balance was soft.

In 1888 E. Golay patented a non-magnetic balance ; the interior ring being formed of an alloy composed of about 40 per cent. platinum, 35 per cent. copper, and 25 per cent. nickel, the external ring consisting of about 55 per cent. silver, 35 per cent. zinc, and 10 per cent. copper.

Mr. T. D. Wright, who found constructional difficulty with brass and zinc, has used with success balances having the arm an inner rim of platinum and the outer part of the rim of silver. He describes them as compensating well, but they required careful handling by reason of their softness.

Good results have also been obtained by the Waltham Watch Company, but the nature of their alloy is not divulged. Messrs. Nicole, Nielsen and Co. have introduced very excellent but costly balances in which the inner part is an alloy of platinum and iridium. Mr. Daniel Buckney patented the use of steel alloyed with about 24 per cent. of manganese for the inner ring. Both the platinum-iridium and the manganese-steel alloys are, I believe, very intractable when subjected to cutters in course of manufacture.

Dealing with the metals and alloys enumerated, it seems that so far as a material for balance springs is concerned, Paillard's palladium alloy appears to best meet all requirements. It is not liable to rust. It is non-magnetic. It is lasting. With it the troublesome "middle temperature error" is materially lessened. Special treatment is, however, required. It is heavier than steel, and therefore smaller springs should be used for watches; the overcoil should be raised more above the body of the spring and brought in nearer to the centre. It is comparatively soft, and when its form is to be altered ivory-tipped or other suitable tools should be used, and it must be bent with care. Bending to and fro will quickly spoil a palladium spring. For balances, there is no particular style or material so far as my observation and enquiry go that can be indicated as being absolutely non-magnetic, yet certain in its action and free from objection.

Steel-Nickel Alloys.—The researches of Dr. C. E. Guillaume, at the Sèvres Office of Weights and Measures, have revealed the fact that steel liberally alloyed with nickel produces a compound with a very small co-efficient of expansion. Dr. Guillaume found the co-efficient of expansion of steel alloyed with 36·2 per cent. of nickel (named Sèvres alloy) to be but 8, in comparison with brass 189; steel or iron, 108 to 122; glass, 86; wood (*bois de sapin*), 44.

Professor M. Thury conducted some experiments with a view of determining the variations in the elasticity of Sèvres alloy when subjected to changes of temperature; and he discovered that in a range of temperature of 22° Centigrade (+ 15° to + 37° Cent.), the elasticity actually increased with a rise in temperature, thus reversing the behaviour of steel under similar circumstances. Such a startling result leads at once to conjectures of a revolution in the methods of compensating watches. Accepting the Sèvres alloy with its low co-efficient of expansion as a suitable material for the balance of a watch, it is easy to suppose that by some slight variation in the proportion of nickel employed a substance may be obtained for the balance spring with a ratio of expansion calculated to entirely neutralize the temperature error in timekeepers furnished with a balance. Though watches with nickel-steel balances and springs have been, I believe, tested with encouraging results, further experiments in the preparation of the alloy, and much labour in the production and application of springs, are of course necessary before one can say positively the position steel-nickel alloys will take in the construction of horological mechanism. Though the steel-nickel alloys cannot be fire-hardened, they are readily compressed and stiffened by hammering or rolling. They are less affected by magnetism than steel, but can hardly be classed under the head of non-magnetic.

CHAPTER VII.
GAUGES.

Hewitt's Balance Spring Tester.—This gauge, designed by Mr. T. Hewitt, is invaluable to the springer of

Fig. 70.

high-class timekeepers. Dr. Hooke, the inventor of the balance spring, in explanation of its isochronous property gave utterance to the now well-worn maxim, *ut tensio sic vis.* (As the tension is, so is the force.) And in theory this is true of springs perfect in construction, though it sometimes happens that a spring apparently faultless contains some latent defect that renders its action not regularly progressive. With the tool shown in the engraving, the behaviour of a spring under different degrees of tension may be tested before it is applied to the timekeeper, and if it prove to be bad it may be at once rejected, and much labour saved that would be otherwise thrown away. The eye of the spring is held in a spring collet, and its outer end caught in a split nose at the end of the arm in front. This arm is held in a split bearing, and is capable of adjustment to suit springs of various kinds. The circular scale is divided into degrees and into radians, and turns friction-tight on its axis so that it may be readily set to zero. By means of the knurled knob at the back of the tool the spring may be wound up, weights being gradually added to the scale pan to balance it. A fair average of the extremes of winding a balance spring would be subjected to in the long and short vibrations of a watch, would be from three to four radians, and if for every tenth of a radian between these points the same increment of weight were required, the progression of the force of the spring under trial would be regular. The bending moment of any spring may be ascertained, and in the case of substituting one balance for another in order to obtain increased vibration, the relative moments of inertia of the old and the new balance may be readily found. The central staff holding the eye of the spring is very nicely fitted into jewelled holes. It carries the lever to which the scale pan is hung at exactly 1 inch from the centre of motion, and then the bending moments, etc., are very readily calculated in inch-grains.

Daldorph's Gauge.—A gauge by Mr. Daldorph which is shown in the accompanying engravings, is a useful form when it is necessary to substitute for a spring another of a different size or strength.

Fig. 71 is a view of the back of the gauge. Upon a plate of brass is mounted an arbor, with a balance spring of medium strength pinned in as shown. The projecting nib

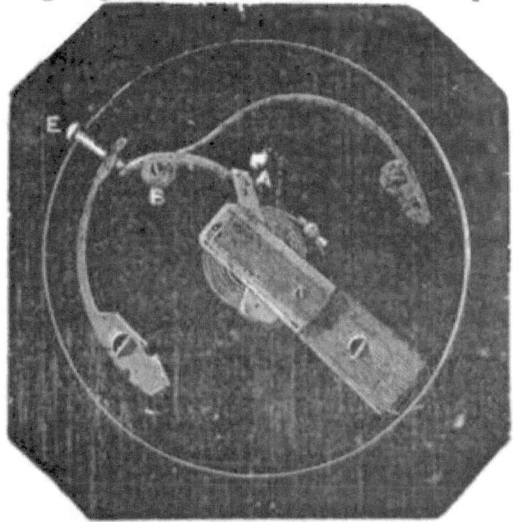

Fig. 71.

A is carried round from its point of rest (that is, bearing

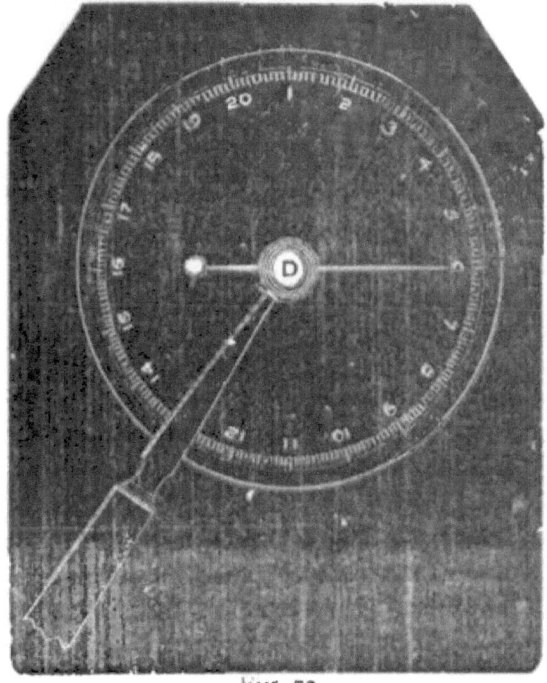

Fig. 72.

against the stop, as shown by the dotted lines) nearly a

whole turn, when it is prevented from flying back by the catch B. Referring to Fig. 72 it will be seen that the arbor is carried through the plate, and has a pointer, resembling a long seconds hand, attached to it. A word of explanation is now required respecting the little holder (D), represented in Fig. 73. The tweezer-like points are kept together by

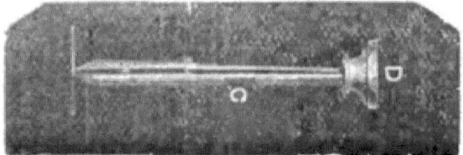

Fig. 73.

the spring forming one of the legs, with just sufficient force to grasp a balance-spring tightly. But if both legs are pressed together somewhere about C the points will open. With this tool take hold of the eye or inner coil of the balance-spring that has to be discarded. The stout leg of the tool has a hole drilled up it to fit the end of the arbor, and, holding the outer coil with an ordinary pair of tweezers (see Fig. 72), push the knob E. The nib A being released the pointer flies back some distance according to the strength of the spring being tried.

The spot indicated by the pointer being noted, the relative strength of any spring which it may be proposed to substitute for the one discarded may be ascertained with great exactness. The whole operation occupies but a remarkably short space of time, and, in addition to the dial being sub-divided into small divisions, it will be a great practical help if the spot indicated by the pointer for every number of springs in general use be marked by corresponding numbers on the dial.

In factories where thousands of watches are produced, each one the counterpart of the others, a more systematic procedure may be adopted than is otherwise possible. The balance springs are all fitted to collets and studs, and their power of resistance measured in the following way: The spring to be tested is held on a centre by a collet, while its stud end is caused to rotate one turn. But in moving from the zero point it had to overcome the resistance of a strong balance spring fixed to the underside of a table, at the edge of which is a finely graduated and numbered scale, and the number of degrees or graduations which the spring under trial is able to move the stronger one attached to the machine is thus registered. The spring is then turned back two rotations, thus testing its resistance to one turn of unbending. The mean of the two readings is then taken, and the girl who presides at the machine takes off the spring tested, and drops it into a compartment marked with a corresponding number. All the balances used are weighed with great exactness, and placed in marked compartments in the same way as the springs. As it is known that a balance of a certain size and weight takes a certain number of spring, the selection of either is thus reduced to a mere perfunctory operation.

CHAPTER VIII.
Observatory Tests.

Kew Observatory.—At the Observatory, Richmond, watches are tested, and if the performance is satisfactory, certificates are issued. The trial for Class A certificates is divided into eight periods of five days each: 1st, pendant up; 2nd, pendant right; 3rd, pendant left; 4th, dial up in refrigerator (about 40° Fahr.); 5th, dial up in room (about 65° Fahr.); 6th, dial up in oven (about 90° Fahr.); 7th, dial down in room; 8th, pendant up. All but periods 4 and 6 are at the ordinary temperature of the room. The 4th, 5th, 6th and 7th periods are each extended one day, and on the first day of each the watch is not rated. To obtain the certificate, the daily rate must not exceed 10 secs.; the *mean difference* of daily rate during each period must not exceed 2 secs.; the difference of mean daily rate between pendant up and dial up must not exceed 5 secs., and between pendant up and any other position, 10 secs.; change of temperature must affect the daily rate by less than one-third of a second per degree Fahrenheit. The behaviour of a watch keeping *just within* the limits throughout the trial is appended.

Mean daily rate	..	..	..	..	.. gaining	9·9 secs.
,, variation of daily rate		..	..	..	..	1·9 ,,
,, ,, ,,	for 1° F.	..	..	..	..	0·3 ,,
Difference of mean rate between pendant up and dial up..						4·9 ,,
,, ,, ,,	,,	pendant right..				9·9 ,,
,, ,, ,,	,,	pendant left	..			9·9 ,,
,, ,, ,,	dial up and dial down..					9·9 ,,

It will be observed that the MEAN variation is taken into account. For instance, between pendant up and dial up the piece may vary more than 5 secs. between one day and the next, provided the MEAN daily variation of the pendant up period does not differ from the MEAN daily rate of the dial up period by more than 5 secs. Marks, showing the excellence of the watch, are awarded to A certificates: 100 marks representing absolute perfection, that is, 40 for no variation of rate, 40 for no change of rate with change of position, and 20 for no temperature error. A watch just touching the limit of variation allowed would get no marks, and for every mark obtained the 2 secs. limit would have to be reduced ·05, the 10 secs. limit ·25, and the temperature error ·015. The words "*especially good*" are added to an A certificate when the watch obtains not less than 80 marks.

For B certificates the watches are tried 14 days pendant up, 14 days dial up, one day in the oven, one day at temperature of the room, and one day in the refrigerator; variation limit in each position 2 secs., and between hanging and lying 10 secs. "*Especially good*" when the 2 secs. limit is reduced to 0·75 sec., the 10 secs. to 5 secs., and the temperature limit 0·2 per sec. Fahr. With the sanction of the Kew Committee, Dr. Charles Chree, superintendent of the Observatory, has favoured me with the following detailed account of the course pursued, and, as an example, the performance of a watch throughout the trial for an A certificate.

On arrival at the Observatory, the watches are wound, divided into classes A and B in accordance with the instructions given by the sender on the " Application Form," and full particulars of the description and escapement, &c., of each watch are entered in the Observatory Register. Each movement then receives a " K.O. register number," by which number it is known and distinguished throughout the trial.

Watches that come to hand a few days previous to the commencement of the trial are kept dial up and wound daily until the test begins. The following short description of the procedure followed refers to the watches entered for the Class A trial, as they form by far the largest proportion of the movements sent for rating :—

When the movements are classified and numbered they are placed upright, with the pendant up, in mahogany racks with semi-circular recesses, specially made for the purpose. The racks are of various sizes so as to accommodate all ordinary watches, and wood or paper pluggets are used to fix any watches not quite secure in their place.

These racks fit into mahogany trays, which run in grooves cut in the sides of the interior casing of a large Milner safe, and are arranged so that each tray can be removed for examination without disturbing neighbouring watches. The interior of the safe is kept as closely as possible at a temperature of 65° F., and to allow this to be done during the colder months of the year, the safe is built up so as to enable a Bunsen gas burner to be fitted beneath it. This burner is provided with a guard shield between the flame and the bottom of the safe, and enables the mean temperature to be regularly maintained. The extremes of temperature variation are registered by maximum and minimum thermometers.

The separate sets of watches are carefully compared at the same time each day [immediately after the noon observations] with the standard mean-time clock, and, as soon as the comparison is made and the reading entered in the Register, the watch is wound up and returned to its place. The error on G.M.T. of the standard clock (always very small) is determined mainly by direct time current from Greenwich, and also—should it be required—by sun and star transits, and by inter-comparisons with two other regulators.

After being observed for six days with the pendant *up*, the watches are placed with the pendant *right* and compared for five days, and then changed to pendant *left*, and the same procedure gone through.

The daily rates on the "quarters" being thus determined, the movements are next tested for temperature compensation, for which purpose they are kept, dial up, for periods of six days each at temperatures of about $40°$, $65°$, and $90°$ F.

The "cold" chamber used for the test at $40°$ consists essentially of a small safe, fitted with a sheet-zinc covering. Built round this a yellow deal casing, surrounded with four inches of non-conducting packing.

The low temperature is maintained by means of blocks of ice, and proper provision is made for draining away the water resulting from the melting of the ice, so that practically the chamber is surrounded by cold air only. The interior is furnished with sliding lattice shelves to hold the watches, dial up, with dishes containing potassic chloride for drying the air, and with thermometers to register the daily range of temperature. To avoid the necessity of taking out the watches into a warmer air during their comparison, the cold chamber is built as near as possible to the clock dial, so that the observer can read and wind the watches without removal.

Experience with some thousands of movements has shown that no danger from moisture condensation during the temperature trials need be apprehended.

The "oven" or hot chamber is designed on similar lines to the "cold" safe, with zinc casing, non-conducting packing, &c., and hot water is employed to keep it at about $90°$ F., the temperature of the water being maintained by means of a Fletcher tubular boiler.

On the completion of the temperature tests, the watches are placed in the large safe (kept at $65°$), dial down, in which position five daily rates are secured, and then they are changed to the final position, which is the same as the first, viz., pendant up.

Having now obtained five daily rates in each of the positions—pendant up, right, and left, dial up and dial down at $65°$, and also with dial up at $40°$ and $90°$ F.—we have the necessary data to give us the marks for positional adjustment, for temperature compensation, and for variation of daily rate.

The accompanying "Rate-record" gives the full history of the performance of a watch entered for the Class A trial, with the average daily rates, variation of rates, and temperatures.

In the "Abstract of Results," supplied with each certificate, these averages are all given, and in addition "Marks" are added for superior performance, based upon the scale 0—100, where 0 represents the marks which would be awarded to the watch which only just managed to pass each separate part of the tests and obtain the bare certificate, while 100 would be the marks gained by the *ideal* and absolutely perfect watch.

The total, which of course is only theoretically possible, is made up as follows :—

For complete absence of variation of rate .... = 40
„ Absolute freedom from change of rate, with change of position .... = 40
„ Perfect temperature compensation .... = 20
                                                    ———
                                                    100

Copy of the "Rate-Records" of gold keyless crystal watch, No. 10001, by X.Y.Z., rated from May 1 to June 14, 1898, and awarded a Class A certificate, especially good :—

In awarding the marks the following rules are followed :

I.— *Variation of daily rate.*

Number of marks possible .... = 40

Limit of variation of rate allowed .. = 2 seconds.

For the number x, of marks awarded we have the formula $x = \dfrac{2 \cdot 0 - V'}{0 \cdot 05}$ where V' is the mean variation of rate, in seconds, for the watch.

Days of Trial.	Posit'n.	Mean Temperature for period.	Daily Rate.	Mean Daily Rate for period.	Variation from Mean Rate.	Mean Variation of Rate.
		Degrees	Secs. *	Secs.	Secs.	Secs.
1—6	I Pend'nt up.	65	+ 1·5		0·00	
		64	+ 0·75		·75	+
		66	+ 1·25	+ 1·50	·25	— 0·40
		64	+ 2·0		·50	
		66	+ 2·0		0·50	
7—11	II Pend'nt right.	67	+ 1·5		0·60	
		66	+ 0·75		·15	+
		66	+ 0·5	+ 0·90	·40	— 0·28
		64	+ 0·75		·15	
		63	+ 1·0		0·10	
12—16	Pend'nt left. III	63	+ 2·25		0·45	
		64	+ 2·25		·45	+
		64	+ 1·5	+ 1·80	·30	— 0·36
		65	+ 1·75		·05	
		65	+ 1·25		0·55	
		INTERMEDIATE DAY.				
17—22	Dial up in Refrigerator. IV.	41	+ 0·25		0·35	
		40	+ 0·25		·35	+
		39	+ 1·0	+ 0·60	·40	— 0·32
		39	+ 1·0		·40	
		41	+ 0·5		0·10	
		INTERMEDIATE DAY.				
23—28	V Dial up at Ordinary Temp.	66	+ 1·75		0·15	
		64	+ 1·75		·15	+
		64	+ 1·5	+ 1·60	·10	— 0·18
		65	+ 1·25		·35	
		65	+ 1·75		0·15	
		INTERMEDIATE DAY.				
29—34	VI Dial up in Oven.	90	+ 2·75		0·60	
		90	+ 2·25		·10	+
		92	+ 2·0	+ 2·15	·15	— 0·28
		90	+ 2·0		·15	
		89	+ 1·75		0·40	
		INTERMEDIATE DAY.				
35—40	VII Dial Down.	64	+ 1·0		0·30	
		65	+ 1·0		·30	+
		66	+ 0·25	+ 0·70	·45	— 0·26
		67	+ 0·75		·05	
		65	+ 0·5		0·20	
41—45	Pend'nt up. VIII	64	+ 1·0		0·30	
		63	+ 0·75		·55	+
		63	+ 1·25	+ 1·30	·05	— 0·36
		65	+ 1·5		·20	
		64	+ 2·0		0·70	

* + Signifies rate gaining. — Rate losing.

In the example of the performance of a watch given on page 126, the average variation of rate in each of the 8 periods of the trial is 0·40, 0·28, 0·36, 0·32, 0·18, 0·28, 0·26, 0·36 second, and the mean of the 8 values $= \pm 0·305$, which corresponds to V' in the above formula. Hence, in this example, the marks awarded in respect of variation of daily rate would be

$$x_1 = \frac{2·0 - 0·305}{0·05} = \frac{1·695}{0·05} = 33·9.$$

II.—Positional Adjustment.

The number of marks possible .. $= 40$

,, limit allowed for positional error $= 10$ seconds.

For the number x_2 of marks awarded we have the formula $x_2 = \frac{10 - p}{0·25}$, where p is the mean change of rate in seconds due to changes of position.

Referring again to the record of watch 10001, we see that the rates in the various positions are:—

    I. Pendant up $= + 1·4$ secs. (mean of periods 1 and 8.)
    II.   ,,   right $= + 0·9$,,
    III.   ,,   left $= + 1·8$,,
    V. Dial up (ord.) $= + 1·6$,,
    VII. Dial down $= + 0·7$,,

The average rate was thus $+ 1·28$ secs., and the rates in the various positions differed from the average as follows:—

    I.   0.12 sec.
    II.   0·38 ,,
    III.   0·52 ,,
    V.   0.32 ,,
    VII.   0·58 ,, giving a mean difference

of $\frac{1.92}{5} = 0.38$ sec., (to the nearest ·01). Hence, applying the formula, we have

$$x_2 = \frac{10.0 - 0.38}{0.25} = \frac{9.62}{0.25} = 38.5 \text{ marks.}$$

III.—Temperature Compensation.

In the case of the temperature compensation, we have two factors to deal with (1) the three different mean temperatures in the trials in the "dial up" position, and (2) the three corresponding mean rates. We first take the mean of the above three mean temperatures, and the mean of the above three mean rates. We next find the departures of the three mean temperatures from their mean, and the departures of the three mean rates from their mean. We then find the value of a quantity, θ, such that

$$\theta = \frac{\text{Sum of departures of the 3 mean rates from their mean}}{\text{Sum of departures of the 3 mean temps. from their mean}},$$

and deduce the marks, x_3, from the formula

$$x_3 = \frac{0.30 - \theta}{.015}$$

This gives a total possible of 20 marks, and allows of an extreme variation of 0·30 seconds for 1° F.

In the case of watch [10001] the three mean temperatures (to the nearest whole degree) are 40°, 65°, and 90°, and the corresponding mean rates are $+0.60$, $+1.60$, and $+2.15$ seconds. The mean of the three mean temperatures is 65°, and the departures from this are 25, 0 and 25, whose sum is 50. The mean of the three mean rates is $+1.45$ secs., and the departures from this are 0·85, 0·15, and 0·75, whose sum (irrespective of sign) is 1·75. Hence in this case

$$\theta = \frac{1.75}{50} = .035;$$

and so by the formula $x_3 = \frac{0.30 - .035}{.015} = \frac{.265}{.015} = 17.7.$

Hence the total marks awarded to this watch would be

In respect of variation of daily rate	....	=	33·9
,, ,, ,, change of rate with change of position		=	38·5
,, ,, ,, temperature compensation	....	=	17·7
			90·1

In the case of *chronograph* watches, at the conclusion of the ordinary Class A trial just described, the running and adjustment of the chronograph work is examined.

The chronographs are run for periods of 24 hours each with the fly-back hands in action (both ordinary and split-seconds), and the effect upon the ordinary daily rate of the watch of this additional work is noted, a change of ± 5 seconds daily being allowed.

Examination is also made of the general behaviour with respect to stopping and starting, action of minute recorder, and freedom of the "fifths" hands from a tendency to "hold-up" with improper use of the plunger, and should any defects be exhibited (now happily much less rare than formerly) the watch is not granted a certificate until the action has been corrected.

"Non-Magnetic" Test.

Watches are subjected to a "non-magnetic" test only when a special request is made to that effect, and when the test has been applied the fact is specifically stated on the certificate.

The examination is conducted with the aid of a powerful electro-magnet, possessing an iron core and moveable pole-pieces, each $2\frac{3}{8}$ inches in diameter.

The watch is placed in the air-gap between the core and pole-piece, **and** its behaviour examined **with** the lines of force passing from front to back, and from **side to** side. In both positions, after being held stationary **for** some time, the watch is slowly rotated through a **complete** revolution, first in one direction **and then in the other.** The rate is noted before and after the magnetic trial, and any serious **change of rate** is made a cause **of** rejection.

Greenwich Observatory.—Here the tests are organized for a different purpose. Watches and chronometers required for the British Navy are purchased after estimation of their value from their performance at the Greenwich trials. Instead of giving marks for good going, a "trial number" is deduced by deviations from exactness. Trials of "Deck Watches" submitted are held annually, beginning in October and lasting 16 weeks. Any manufacturer is accorded permission to deposit suitable timekeepers on written application to the Hydrographer to the Navy at Whitehall. The watches are used principally for noting the time when observations of the sun or stars are being taken. The chronometers not being on deck their error is ascertained by comparison with the deck watches. Watches for this purpose are preferred to be of about No. 20 size, with seconds dial not less than $\frac{1}{17}$ of an inch diameter, accurately and distinctly divided. The hour and minute hands and the hour figures should not be so heavy as to interfere with **the** visibility of the seconds.

The trial is conducted in the following way:—

Watch	horizontal,	dial up,	in room	for	6 weeks
,,	,,	,,	in oven	,,	1 week
,,	vertical, pendant up		,,	,,	4 days
,,	,,	,,	right	,,	,,	3 ,,
,,	,,	,,	left	,,	,,	3 ,,
,,	,,	,,	up	,,	,,	4 ,,
,,	horizontal, dial up			,,	,,	1 week
,,	,,	,,	in room	,,	,,	6 weeks
		Total duration of trial		..	16 weeks

The mean temperature in the oven is from 80° to 90° Fahrenheit.

The watches are wound and noted every morning but to obtain the trial number the rating is summarized in this way:

Difference between greatest and least weekly rates. Dial up. *a*	Greatest difference between one week and the next. Dial up. *b*	Difference of Weekly Rates.			Trial Number. $a + 2b\left(\dfrac{c+d+e}{3}\right)$
		Pendant up —Dial up. *c.*	Pendant left. —Dial up. *d.*	Pendant left. —Dial up. *e.*	
s. 14·8	s. 4·4	s. +1·2	s. −4·0	s. −4·5	25·8

The figures given are those relating to the first watch on the 1898 list. It will be observed that in fomulating the trial number steady going is regarded as of primary importance, and errors between any week and the next carry double the penalty of deviations spread over the 16 weeks of trial. The trial number obtained in the case quoted (25·8) is small. The greater the variations the higher the trial number, and in preparing the list watches with a higher trial number than 100 are unnoticed.

Marine chronometer trials are also conducted annually at Greenwich. They begin in July and extend over 29 weeks. They are tested in the oven for two periods of four weeks each, at temperatures ranging from 75° to 100° Fahrenheit, and also at the ordinary summer and winter temperatures of the room.

To obtain the trial number the rates are summarized as below. The figures given represent the performance of the chronometer occupying the premier position in the 1898 Report.

Difference between greatest and least weekly rates. *a.*	Greatest difference between one week and the next. *b.*	Trial Number. *a* plus 2*b*.
s. 6·0	s. 2·8	11·6

Note on Timing Repeating Carriage Clocks.—
The quickest method is to listen to the first blow of the hammer on the bell (or gong) at each hour and half-hour, noting at the same time the position of the seconds hands on the regulator. Thus: supposing the blow is given exactly at **12 h. 0 m. 0 s.** make a note of it, and compare again at **12 h. 30 m. 0 s.**, when the half-hour blow is given at, say, 12 h. 30 m. 5 s., make a note of this also, but do not alter the index, as this difference may arise from the imperfect position of the half-hour pin in the cannon wheel; at 1 h. 0. m. 0. s. compare as before, and as it is the same pin coming into action as at 12 h. 0 m. 0 s., any difference in the time of the first blow will be indicative of a gain or loss, as the case may be, thus: it may be given at 1 h. 0 m. 4 sec., which will show a loss of 4 s. in the last hour, and the clock may now be regulated accordingly. The blow may next be given at 1 h. 30 m. 8 s., showing the effect the moving of the index has had in making it gain 1 s. in the preceding half-hour; regulate and compare again at **2 h. 0 m. 0 s.**, and if the first blow is given at 2 h. 0 m. 3 s. it will show that the clock has been keeping correct time for the last half-hour.

CHAPTER IX.

EXAMINATION OF ESCAPEMENTS.

Examination of the Lever Escapement.—See that the balance staff is perfectly upright. See that the escape wheel is perfectly true on edge and on face, and that the teeth are equally divided and smooth; also, by gently turning the wheel backwards, see that the pallets free the backs of the teeth. If the wheel is out of truth it must be set up in the lathe and re-bored. It can be fixed either with shellac or in a brass sink bored out the exact size to receive it. If the divisions are unequal, or the wheel has some thick teeth, it should be discarded. It is useless to attempt to make the wheel right, and to reduce the corners of the pallet to free the wheel is simply to spoil the escapement for the sake of the wheel. At the same time, it must be left to the operator to judge whether the amount of the inaccuracy is serious. The whole affair is so minute that no rule can be given.

Is the wheel the right size? If the lockings are too light, and the greater part of the shake INSIDE, the wheel is too small, and should be replaced by one larger. Before removing the wheel, gently draw the balance round till the point of the tooth is exactly on the locking corner, and see if there is sufficient shake. If not, it will be prudent to have the new wheel with the teeth a little straighter than the old ones. If the lockings are too deep and most of the drop OUTSIDE, the wheel is too large and should be topped.*

The wheel is so fragile that care is required in topping, which is done by revolving it in the turns against a diamond or sapphire file. A brass collet is broached to fit friction-tight on one of the runners of a depth tool: one side of this collet is then filed away, leaving sufficient substance to avoid bursting into the hole. On this flat a small piece of sapphire file is attached with shellac, taking care that the *face of the file is parallel to the centre of the runner*. The escape wheel on its pinion with the ferrule attached is placed in the centres of the depth tool *furthest from the adjusting screw*, and the collet and file on one of the opposite centres, and that centre fixed firmly by its clamping screw. A very light hair bow is used to rotate the pinion, and the depth tool laid on its side on the work-board, the tool being closed by its screw until the teeth of the wheel *nearly* touch the surface of the file; now if a slight pressure is made by the fingers on the uppermost limb of the tool, at the same time rotating the wheel by the bow, the *spring*

---

*In planting the wheel and pallets it is always best to err, if at all, by making them too deep rather than too light. If they are a shade deep, topping the wheel soon puts matters right.

of the tool will allow the teeth to be brought into contact very slightly and without fear of bending the teeth; the wheel can be reduced as much as is necessary.

If the wheel is the right size and there is no shake (which try, as before directed), the discharging corner of the pallets may be rounded off by means of a diamond file if they are of garnet. If they are of ruby, they may be held against an ivory mill charged with diamond powder. If the lockings are too light and there is but little shake, they may be made safe by polishing away the locking face a sufficient quantity. If one locking is right and one is too light, the one that is too light may be made safe by polishing away the locking face as before, or the pallet may be warmed and the stone brought out a bit. The locking faces of the pallets should be sufficiently undercut to draw the lever to the banking pins without hesitation. If they require alteration in this respect, polish away the upper part of the locking faces so as to give more draw, leaving the locking CORNER quite untouched. But proceed with great care, lest in curing this fault the watch sets on the locking, as small watches with light balances are very liable to do. If a watch sets on the lockings, or on one of them, the locking face or faces may be polished away so as to give less draw — *i.e.*, have most taken off the CORNER of the locking. If the watch sets on the impulse, the impulse face may be polished to a less angle if the locking is sufficiently deep to allow it, for it must be remembered that in reducing the impulse the locking of the opposite pallet will also be reduced. In fact, the greatest caution should be exercised in making any alteration in the pallets.

Sometimes, in new escapements, the oil at the escape wheel teeth will be found to thicken rapidly through the pallet cutting the wheel, showing that one or both corners of the pallet are too sharp. If ruby, the corner may be polished off with a peg cut to the shape of a pivot polisher, and with a little of the finest diamond powder in oil; if garnet, diamantine on a peg will do it very well. Great care should be taken to remove every trace of the polishing material, or the wheel may become charged with it.

See that the pivots are well polished, of proper length to come through the holes, and neither bull-headed nor taper. A conical pivot should be conical only as far as the shoulder; the part that runs in the hole must be perfectly cylindrical. They must have perceptible and equal side shake, or if any difference be made the pallet pivots should fit the closest. Both balance staff pivots should be of exactly the same size. The end shakes should all be equal. Bad pivots, bad uprighting, excessive and unequal shake of the pivots are responsible for much of the trouble experienced in position timing. With unequal end shakes the pallet depth is liable to be altered owing to the curved form of the pallet faces. The action of the escapement will also be affected if the end shakes are not equal, by a banking pin slightly bent, a slight inaccuracy in uprighting, and other minute faults. The infinitesimal quantity necessary to derange the wheel and pallet action may be gathered from the fact that a difference of ·002 of an inch is quite enough to make a tripping pallet depth safe or correct depth quite unsound.

When the wheel and pallets are right see that the impulse pin is in a line with an arm of the balance, and proceed to try if the lever is fixed in the correct position with relation to the pallets. Gently move the balance round till the tooth drops off the pallet. Observe the position of the balance arm, and see if it comes the same distance on the other side of the pallet hole when the other pallet falls off. If not, the pins connecting pallet and lever are generally light enough to allow of the lever being twisted. To do this successfully a clamp to grasp the back and belly of the pallets, as shown in Fig. 74, or some similar tool is necessary.

Mr. R. Bridgman has devised a very superior clamping tool for pallets, which, with some modification by Mr.

Fig. 74.

C. Curzon, is shown in Fig. 75. There is a spring for keeping the slide in contact with the pallets when they are placed in position, and until they are gripped tight by the screw, and a lever under the body of the tool by means of which the spring is overcome and the

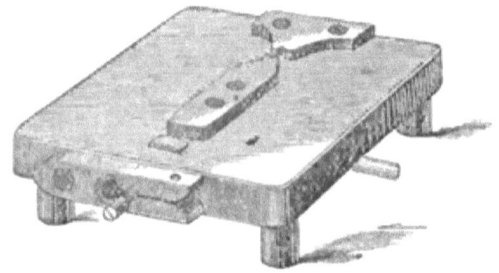

Fig. 75.

pallets released when the operation is completed. It will be seen that the screw passes through a split nut, which may be set up as it wears. The jaws of the tool should be faced with tin to avoid marking the pallets. When the lever is right with relation to the pallets, see that the pallets are quite firmly fixed to the lever, and that the lever and pallets are perfectly in poise. This latter is an essential point in a fine watch to be timed in positions, but it is often neglected.

See that the escapement is in beat. When the balance spring is at rest, the impulse pin should be on the line of centres, that is in the middle of its motion. If this is not so, the spring should be drawn through or let out from the stud if the position of the index allows; if it does not, the roller may be twisted round on the staff in the direction required.

Is the roller depth right? If the safety pin has insufficient freedom while there is enough run, the roller is probably planted too deep. On the other hand, if it is found that while the safety pin has plenty of freedom there is no shake between the bankings the roller depth is probably too shallow. When the impulse pin is led round there should be an equal clearance all round the inside of the horn, and the pin must fall safely into the notch. If it binds in the horn and bottoms in the notch it is too deep, and, on the other hand, if with excessive clearance in the horn the pin when it falls does not pass well into the notch it is too shallow. The readiest method of altering is to warm the roller, remove the impulse pin, and using a to-and-fro motion with a wire and oilstone dust, draw the hole in the required direction. If the pin is deep in the notch and too tight in the roller to give a little, it should be removed and flattened off a trifle more. If too shallow, a triangular pin, or one of some other shape with the point of contact more forward, can generally be substituted by polishing out the hole towards the crescent. If not, the staff hole in the lever may be drawn to allow of shifting the lever sufficiently; or the recesses for the jewel settings of the balance staff pivots may be scraped away on one side and rubbed over on the other to suit. See as it passes round that the impulse pin is free when in the notch.

Just as the safety pin is about to enter the crescent, the impulse pin must be well inside of the horn. In the single roller escapement a very little horn is required, unless the crescent has been made of an unnecessary width. In very common work one occasionally sees a flat filed on the edge of the roller instead of a crescent. There is no excuse for such a piece of bungling.

A fault occasionally met with is that the impulse pin after leaving the notch just touches on some part of the inside of the horn in passing out. If a wedge of cork is placed under the lever, so that the lever moves stiffly, it can be readily seen whether or not the impulse pin is free to leave the notch, and is free all round the horn when the wheel tooth drops on the locking.

See to the safety action. When the tooth drops on to the locking, the safety pin should be just clear of the roller. If it is not clear, the edge of the roller should be polished down till it is right. If there is more than clearance, the safety pin must be brought closer to the roller. See upon pressing the safety pin against the roller that the tooth does not leave the locking, and that the impulse pin is free to enter the notch without butting on the horn of the lever: also that the safety action is sound, so that the pin is in no danger of passing the roller. If the action is not sound the diameter of the roller should be reduced and the safety pin brought towards it sufficiently to get a sound action if it can be done, but if the escapement has been so badly proportioned as not to allow of a sound action being obtained in this way, the pin must be shifted forward and the bankings opened to allow more run.

See if the banking pins are so placed as to allow of an equal run on each side. If not, they should not be bent, for with bent banking pins a difference in the end shakes of the pivots will cause a difference in the run. The banking pin allowing of the most run should be removed, and the hole broached out to receive a larger pin.

A fault rather difficult to detect, which is sometimes met with in the double roller escapement, is the end of the impulse pin slightly touching the safety finger, caused by excessive end-shake, or from a longer impulse pin than was originally intended having been put in.

To Examine the Chronometer Escapement.— See that the wheel is true and the teeth smooth and perfect, and that the rollers properly fit the staff. See that the end shakes and side shakes are correct. See that that the "lights" between the wheel teeth and the edge of the roller are equal on both sides when the wheel is locked. If they are not, the foot of the detent must be knocked a trifle to or from the centre of the roller till the lights are equal. If the light is more than sufficient for clearance, the roller must be warmed to soften the shellac, and the impulse pallet moved out a little. If the light is excessive there will be too much drop on to the locking after the wheel tooth leaves the impulse pallet, and with a large drop there is danger of tripping.

To ensure safe locking the detent should be set on so that when the banking screw is removed, and the locking pallet is free of the wheel teeth, it will just spring in as far as the rim of the wheel.

In pocket chronometer escapements it is especially necessary to see that the face of the locking stone is angled so as to give perceptible draw. Many pocket chronometers fail for want of it.

The gold spring should point to the centre of the roller. Bring the balance round till the discharging pallet touches the gold spring preparatory to unlocking, and notice how far from that point the balance moves before the gold spring drops off the face of the pallet. Then reverse the motion of the balance, and see if the same arc is travelled through from the time the *back* of the pallet touches the gold spring till it releases it. If not, the horn of the detent must be bent to make the action equal.

Bring the discharging pallet on to the gold spring, and let it bend the detent so that the locking stone is as much outside the wheel as it was within when the wheel was locked. The gold spring should then drop off the discharging pallet. Make it to length, sloping off the end from the side on which the pallet falls to unlock, and finish it with great care. The gold spring should be thinned near its fixed end as much as possible, and the detent spring thinned if it is needed. The judgment of the operator must determine the proper strength in both cases. The nose of the detent horn should be nicely flattened and the corners rounded off.

The locking pallet should not be perfectly upright. It should lean a little from the centre of the wheel, and a little towards the foot of the detent, so that the locking takes place at the root of the stone, and then the action of locking and unlocking does not tend so much to buckle the detent. The face of the impulse pallet, too, should be slightly inclined, so that it bears on the upper part of the wheel teeth. By this means the impulse pallet will not mark the wheel in the same spot as the locking pallet.

Try if the escape wheel teeth drop safely on the impulse pallet by letting each tooth in succession drop on, and, after it has dropped, turn the balance gently backwards; you can then judge if it is safe by the amount the balance has to be turned back before the tooth leaves the pallet. If some teeth do not get a safe hold, the impulse roller must be twisted round on the arbor to give more drop.

If the escapement is in beat, the balance, when the balance spring is at rest, will have to be turned round an equal distance each way to start the escapement. When the balance spring is in repose, the back of the discharging pallet will be near the gold spring, and if the balance is moved round till the gold spring falls off the back of the pallet and then released, the escapement should start of itself; and in the other direction also, if the balance is released directly the wheel tooth leaves the face of the impulse pallet, the escapement should go on of itself.

Examination of the Cylinder Escapement.— See that cylinder and wheel are perfectly upright. Put the cylinder and cock in their places. Then, with a little power on, and a wedge of cork under the balance to check its motion, try if all the escape wheel teeth have sufficient drop, both inside and out. If, with the cylinder planted the correct depth, there is sufficient drop inside the cylinder and none without, the cylinder is too large; if the reverse fault is apparent, the cylinder is too small. If some of the teeth only are without necessary freedom, make a hole in thin sheet brass of such a size that one of the teeth that has proper shake will just enter. Use this as a gauge to shorten the full teeth by. For this purpose use either steel and oilstone dust or a sapphire file, polish well with bell metal and red stuff, and finish with a burnisher. Be careful to operate on the noses of the teeth only, and round them both ways so that a mere point is in contact with the cylinder. If the inside drop is right, and there is no outside drop with any of the teeth, the cylinder may be changed for one a little smaller or for one of the same inside diameter, but thinner. Or the wheel may be changed for one a little LARGER, but in this case be sure the larger wheel will clear the fourth pinion. And with insufficient drop inside changing the wheel for one a little smaller will often be more expeditious than removing the cylinder.

If the teeth of the escape wheel are too high or too low in passing the opening of the cylinder, the wheel should be placed on a cylinder of soft brass or zinc small enough to go inside the teeth, with a hole through it and with a slightly concave face. A hollow punch is placed over the middle of the wheel while it is resting on the concave face of the brass or zinc cylinder, and one or two light taps with a hammer will bend the wheel sufficiently. In fact, care must be taken not to overdo it. It rarely happens that the wheel is free neither of the top nor bottom plug, but should this be the case sufficient clearance may be obtained by deepening the opening with a steel polisher and oilstone dust or with a sapphire file. A cylinder with too high an opening is bad, for the oil is drawn away from the teeth of the escape wheel.

If a cylinder pivot is bent, it may very readily be straightened if a *bouchon* of a proper size is placed over it to get a leverage.

When the balance spring is at rest, the balance should have to be moved an equal amount each way before a tooth escapes. By gently pressing against the fourth wheel with a peg this may be tried. There is a dot on the balance and three dots on the plate to assist in estimating the amount of lift. When the balance spring is at rest, the dot on the balance should be opposite to the centre dot on the plate. The escapement will then be in beat—that is, provided the dots are properly placed, which should be tested. Turn the balance from its point of rest till a tooth just drops, and note the position of the dot on the balance with reference to one of the outer dots on the plate. Turn the balance in the opposite direction till a tooth drops again, and if the dot on the balance is then in the same position with reference to the other outer dot the escapement will be in beat. The two outer dots should mark the extent of the lifting, and the dot on the balance would then be coincident with them as the teeth dropped when tried this way; but the dots may be a little too wide or too close, and it will therefore be sufficient if the dot on the balance bears the same *relative* position to them as just explained; but if it is found that the lift is unequal from the point of rest, the balance spring collet must be shifted in the direction of the least lift till the lift is equal. A new mark should then be made on the balance opposite to the central dot on the plate.

When the balance is at rest, the banking pin in the balance should be opposite to the banking stud in the cock, so as to give equal vibration on both sides. This is important for the following reason: The banking pin allows nearly a turn of vibration, and the shell of the cylinder is but little over half a turn, so that as the outside of the shell gets round towards the centre of the escape wheel, when a tooth is at rest outside of the cylinder, the point of a tooth may escape over the exit lip and jamb the cylinder unless the vibration is pretty equally divided. When the banking is properly adjusted, and a tooth is at rest inside the cylinder, bring the balance round till the banking pin is against the stud; there should then be perceptible shake between the cylinder and the plane of the escape wheel. If there is no shake the wheel may be freed by taking a little off the edge of the passage of the cylinder where it fouls the wheel by means of a sapphire file, or a larger banking pin may be substituted at the judgment of the operator. See that the banking pin and stud are perfectly dry and clean before leaving them: a sticky banking often stops a watch. Cylinder watches and timepieces, after going for a few months, sometimes increase their vibration so much as to persistently bank. To meet this fault a weaker mainspring may be used, or a larger balance, or a wheel with a smaller angle of impulse. By far the quickest and best way is to *very slightly* top the wheel by holding a piece of Arkansas stone against the teeth, afterwards polishing with boxwood and red stuff. So little taken off the wheel in this way as to be hardly perceptible will have great effect.

Revolving Escapements.—The tourbillon, invented by A. L. Bréguet, is a revolving carriage in which the escapement is placed, the object of the revolution being to eliminate the errors due to varying positions, and particularly the quarter positions, which present the greatest difficulty to the adjuster. In its original form, it is scarcely ever applied except to watches subjected to Observatory or other competitive trials.

B. Bonniksen has invented a more compact arrangement which he calls a karrusel, in which the carriage driven by the third pinion rotates once in 52½ minutes, which gives sufficiently quick change of position for all practical purposes.

Bréguet's and Bonniksen's devices are both illustrated in the *Watch and Clockmakers' Handbook*, and I do not think anything need be said here in reference thereto, except a warning that the large pivot carrying the karrusel carriage should not be oiled. The rotation is so very slow that lubrication is unnecessary, and would be mischievous.

INDEX.

Acceleration	16
Adjustable stud	95
Adjusting for positions	32
,, ,, temperature	36
Airy	38
Aluminium bronze	100, 107, 109
Annealing	100
Applying a balance spring	48
Arnold, John	37
Attachment with relation to pendant	32
Auxiliary	43
Balance	4, 86
,, arc	30, 31
,, cock	48, 97
,, compensation	39, 81
,, spring	4
,, applying	48
,, hardening	101
,, spring holder	56
,, spring making	97
,, spring tester	117
,, staff	135
Balances, size of	5, 11
Banking	138, 143
,, pins	138
Barber, W. N.	18, 25, 35, 71, 134
Barrow's Regulator	10
Beat	136
Benzine	105
Berthoud	37
Bickley	55
Bird cage	7
Bluing	105
Boiling-out pan	98
Bonniksen	67, 144

K

Bosley	10
Breguet	15, 143
,, drawing of curve	15
,, spring	15
Bridgman, R.	92, 136
Buckney, D.	115
Callipers	90
Carriage clock	132
Centre of gravity	32, 33
,, ,, gyration	5
Centrifugal tendency	13
Charles II.	8
Chree, Dr. Charles	**122**
Chronometer escapement	32, 141
Clamp for lever pallets	136
Cock	48, 97
Cole, J. F.	14, 107
Collet	49, 52
,, adjuster	55
,, lifter	55
,, table for	97
Compensation adjustment	36, 81
,, balance	38
,, curb	36
Counting vibrations	48
Crescent	137
Cumming, A.	37
Curb pins	51
Curzon, C.	136
Cyanide	97
Cylinder escapement	30, 141
Cylindrical springs	**13**
Daldorph's stud and collet table	97
,, gauge	118
Dead point	29
Deck watches	130
Dennison, B.	13
Dent, E. J.	46
Depthing	73
Derham	7
Detached escapement	32, 139
Detent	**139**

Dial	71
Draw	134
Drilling	74
Drop	31
Duo in uno	6
Duplex escapement	12
Earnshaw, T.	37
Elasticity	5, 6
Emery	37
End-stone	68
Escapement, examination of	132
,, chronometer	139
,, cylinder	141
,, lever	132
Escape pinion	48
,, wheel	133, 140, 141
Escaping arc	30, 31
Escape wheel too heavy	68
Excelsior callipers	92
Expansion of metals	37
Flat spring	6, 48
Fourth wheel	14
,, pinion	48
Free spring	59
Friction	68
Frictional escapement	30, 141
Gardner, R.	6, 14, 89
Gas governor	84
,, ,, Kullberg's	85
,, ,, Mercurial	85
Glasgow	100
Going barrel	60
Golay	115
Gravity	32, 33
Greenwich	36, 47, 130
Grossmann Jules	18, 33
Guard pin	138
Guildhall Museum	43
Guillaume, C. E.	116
Guye's adjustable stud	95

Hammersley	6, 17
Hands	110, 130
Hardening	101
Hardy, W.	45
Harrison, J.	36
Hartnup	42
Heavy escape wheel	68
Helical springs	13
Huygens	8
Hog's bristle	9
Hooke, R.	7, 17, 108, 117
Horizontal escapement	30, 141
Houriet, F.	7, 110
Houston	113
Hydrochloric	102
Ice box	85
Impulse pin	137
Index	10, 137
Inertia	5
Isochronism	17, 61
Isochronous	18, 64
Jewel hole	50
Joint pusher	53
Journal Suisse d'Horlogerie	18
Kew Observatory	121
Karrusel	144
Knudsen	109
Kullberg, V.	15, 84
Kullberg's balance	46
Leroux, John	43
Locking pallet	140
Lever escapement, examination of	132
Line of centres	78
Long arcs, extra defective	68
Loseby, E. T.	46
Lossier, L.	18, 21, 22, 36
Macartney's collet adjuster	55
Magnetism	109
Mainspring, stronger to quicken short arcs	69

Marine chronometer	14, 41, 131
Mass.	5
Massey, E.	46
Mean time	77
Mercer	107, 108
Method of forming terminal curves	25
Middle temperature error	40
Modulus of elasticity	6
Molyneux, R.	44
Measurement	72
Moment of elasticity	5
Moray	7
Movement holder	60
Nickel	111, 112
Nickel-steel	116
Nicole, Nielsen & Co.	115
Non-magnetizable watch	109
North, R. B.	106
Norwegian refrigerator	81
Oil.	42
Oven	82
,, Hearson's	82
,, Schoof's	83
Overcoil	6, 11, 19
Paillard	110
Palladium springs	33, 42, 110, 115
Pallet	30, 31
,, staff	30
Passing spring (gold spring)	139
Point of attachment	78
Phillip's spring	18, 20
Pinning in, under or over full turns	12, 35
Pivot	135
Plate	50
Platinum	112
Pliers	55
Plose	55, 87
Plose's collet lifter	55
Pocket chronometer	14, 32
Point of attachment to collet	32
Poising	90

Poising tool	91
Polishing	104
Position errors	45
Poole, J.	45
Position errors	63
Quarter screws	63
Rack	60
Rambal	115
Red-stuff	88, 104
Refrigerator	81, 85
Regulator	49
Roberts	9
Remedies	67
Run	137
Rust	109
Safety pin	137
Sapphire	141
Schloss	10
Schoof	83
Screw	39
,, driver	89
,, left-handed	107
Seconds hand	110
Sevres alloy	116
Short arcs, to quicken	63
Silver	112
Single beat escapement	14
Spherical	7
Springs, making	97
Staff	135
Steel	99
Steel nickel	116
Stud	94
,, adjustable	95
,, remover	97
Table roller	137
Tempering	103
Terminal curves	17
Thermometer	103
Theoretical curves	17

Thury	116
Timing	64
„ carriage clocks	132
„ in positions	64
„ in reverse	64
„ screws	39
Tompion, T.	9
Tourbillon	143
Train	48
Trial number, deck watches and marine chronometers	131
Timing box	60
Turns	71
Tweezers	97
„ for balance springs	57
„ „ studs	97
Ulrich	110
Unlocking resistance,	30, 134
Vibrations	48
Vibrator	51
Villarceau	38
Walker, George	18, 25, 35
Walker and Barber's method of forming and testing terminal curves	25
Walker and Barber's adjustable stud	95
Walsh	40
Waltham Watch Co.	115
Watch tests	121
„ balances	4
„ size of	11
Webb, A.	40, 103
Wilkins	7
Woerd, C. V.	46
Wright, T. D.	38, 115
Young	6

[While these sheets are passing through the press, I learn, with deep grief, the Mr. George Walker, whose name frequently occurs therein, is no more. He was a thorough horologist, devoting attention particularly to the laws which govern the motion of the balance and balance spring, and an expert springer. But two months ago he made the sketches for Walker and Barber's tester, described on pp. 26—29, and had he lived would have contributed an article dealing with the turning of overcoils and other points on which he was admittedly an authority.]

October, 1898. F. J. B.

www.ingramcontent.com/pod-product-compliance
Lightning Source LLC
Chambersburg PA
CBHW030340170426
43202CB00010B/1182